CUTNELL & JOHNSON

PHYSICS
6th Edition

Probeware Lab Manual
Student Version

Exploring Physics
With PASCO Technology

PASCO®

John Wiley & Sons, Inc.

Cover Photo: John Kelly/The Image Bank/Getty Images

To order books or for customer service call 1-800-CALL-WILEY (225-5945).

ISBN 13: 978-0-471-47675-7

10 9 8 7 6 5 4 3 2 1

Contents

Introduction

Letter to Student v

About the Technology vii

Activities List

Mechanics

CJ 01   Motion in One Dimension 1
CJ 02   Position, Velocity, and Acceleration 5
CJ 03A Projectile Motion Part 1 – Change Initial Speed 9
CJ 03B Projectile Motion Part 2 – Change Launch Angle 13
CJ 04A Newton's Second Law Part 1 – Constant Mass 19
CJ 04B Newton's Second Law Part 2 – Constant Net Force 25
CJ 05A Newton's Third Law Part 1 – Collisions 31
CJ 05B Newton's Third Law Part 2 – Tug-of-War 35
CJ 06   Work and Energy 39
CJ 07   Conservation of Energy 45
CJ 08   Impulse v Change in Momentum 51
CJ 09A Conservation of Linear Momentum Part 1 – Elastic Collision 55
CJ 09B Conservation of Linear Momentum Part 2 – Inelastic Collision 59
CJ 10   Rotational Motion 65
CJ 11   Hooke's Law 69
CJ 12   Simple Harmonic Motion, Mass on a Spring 73
CJ 13   Simple Harmonic Motion, Simple Pendulum 77
CJ 14   Buoyant Force 83

Thermodynamics and Waves

CJ 15   Temperature and Heat 87
CJ 16   Specific Heat 93
CJ 17   Ideal Gas Law 99
CJ 18   Superposition 105
CJ 19   Interference in Sound–Beats 111

Electricity and Magnetism

CJ 20   Ohm's Law 117
CJ 21   DC Series Wiring 121
CJ 22   DC Parallel Wiring 127
CJ 23   RC Circuit 133
CJ 24   Magnetic Field Around a Wire 139
CJ 25   Magnetic Field of a Solenoid 143
CJ 26   Faraday's Law 147

Optics

CJ 27   Polarization 151
CJ 28   Diffraction of Light 155

Modern

CJ 29   Spectral Lines 161
CJ 30   Photoelectric Effect – Planck's Constant 167

Letter to Student

Welcome to this innovative, technology – infused physics lab manual. The activities were specifically selected to support your use of Cutnell & Johnson, Physics, 6th Edition. The activities also match topics recommended in the Advanced Placement Physics B Course Description published by The College Board.

The activities are designed for use with PASCO equipment and PASPORT probeware. Each activity has an accompanying DataStudio Workbook (sometimes called an "e-Lab") and DataStudio Configuration file. These electronic files are usually installed inside the folder "eLabs" inside the folder "DataStudio".

Each Workbook is a multimedia electronic workbook that includes background from the text, a materials list, setup instructions, data recording procedures, data displays, analysis suggestions, and extension applications and problems. The Workbook parallels the written lab activity and can be used independently.

Each Configuration file has one or more pre-configured data displays. The Configuration file requires the written lab activity found in this manual but provides the option of performing the exploration outside the structure of the electronic workbook.

How To Start a DataStudio Workbook or Configuration File

DataStudio will automatically start up and configure itself when a PASPORT sensor is connected to the computer. For the simplest operation, connect a PASCO USB Link to the USB port on the computer. Next, connect a PASPORT sensor to the USB Link. The DataStudio monitor program called PASPORTAL will open. The PASPORTAL window lists all the Workbook and Configuration files in the eLabs folder that use the connected PASPORT sensor. (For example, if you connect a Motion Sensor to the computer, the PASPORTAL window lists the Workbook and Configuration files that use the Motion Sensor.)

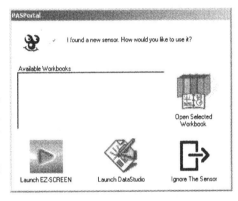

If you select a Workbook in the PASPORTAL window, follow the instructions in the Workbook. If you select a Configuration file, follow the instructions in this lab manual.

If you are working in a group, we suggest that each group member take notes in their science lab notebook.

When completed, your instructor will review your DataStudio Workbook or Configuration file: make sure that it reflects your insights into physics. Make sure that you save your DataStudio e-Labs properly! Your instructor will give you the guidelines.

About the Technology

The hardware and software required to perform these activities were specifically designed by PASCO for education. We understand how important it is to have the technology support the educational process, not require an education in itself.

With PASCO's DataStudio software and PASPORT probeware, you are able to collect real-time data in the lab or in the field. In addition, you are able to select among several modes of data display (e.g. line charts, digits displays, meter displays, tables and bar charts, etc.) and methods of analysis. With electronic workbooks, you can deliver labs electronically or print them out on paper.

DataStudio, an all-in-one software solution, collects, displays, stores, and analyzes scientific data, using either a PC or Macintosh. You can learn by asking and testing "what if . . ." questions that reinforce scientific principles. DataStudio is not a simulation program. It allows you to do real-time data acquisition and analysis. It is easily integrated across scientific disciplines as well as into mathematics classrooms.

The line of PASPORT probeware is designed for true plug-and-play capability. PASPORT probeware is simple and rugged but also precise and reliable enough to be used for college-level experimentation. You can connect any PASPORT sensor into the USB port on a computer using any PASPORT interface (e.g., USB Link, Xplorer Datalogger, or PowerLink).

Last but not least, PASCO's innovative line of physics equipment, specifically designed for physics education, is a staple in physics labs around the world.

Computer requirements:

Macintosh: CD-ROM drive, USB port, Mac OS 8.6 or higher.

PC: CD-ROM drive, USB port, Windows 98 or higher.

You must have DataStudio version 1.7 or higher installed on your computer. (The PASCO website at **www.pasco.com** has information about the latest version of DataStudio.) The DataStudio plugin program called WAVEPORT is required for two of the activities.

Support

If you have any questions about using DataStudio software or PASPORT sensors please contact PASCO technical support at:

(800) 772-8700 (within the US)

(916) 786-3800 (outside the US)

PASCO technical support can be reached on-line at:

techsupp@pasco.com   (email)
http://www.pasco.com (Worldwide Web)

PASCO scientific

Activity 1: Motion in One Dimension
(Motion Sensor)

Preface

- *If* you are using the PASCO electronic Workbook specifically designed for this activity, then do the following:
1. Connect the *USB Link* to the computer's USB port.
2. Connect the *Motion Sensor* to the USB Link. This will automatically launch the PASPORTAL window.
3. Choose the electronic Workbook entitled: **01 Motion in One Dimension WB.ds** and follow the directions in the Workbook.

Introduction

The purpose of this activity is to explore graphs of motion (position versus time and velocity versus time). Use a Motion Sensor to measure your motion as you move back and forth relative to the sensor in a straight line at different speeds.

The challenge is to move in such a way that a plot of your motion will "match" the position-time and velocity-time graphs that are provided for you. Examine your motion using the graphs. The goal is to build an understanding of the relationships between position and velocity.

Learning Outcomes

You will be able to:

- Understand the relationship between position and velocity.

- Understand that velocity is the rate of change of position over time.

- Relate the physical motion of an object to graphical representations of the motion.

- Use the software to determine how well your motion 'matches' the pre-defined plots of position and velocity on the Graph display.

- Understand the idea that all motion occurs within a frame of reference.

Hypothesis

What is the relationship between the motion of an object and the graphs that represent its position versus time and velocity versus time?

Background

When describing the motion of an object, knowing where it is relative to a reference point, how fast and in what direction it is moving, and how it is accelerating (changing its rate of motion) is essential. A sonar ranging device such as the PASPORT Motion Sensor uses pulses of ultrasound that reflect from an object to determine the position of the object. As the object moves, the change in its position is measured many times each second. The change in position from moment to moment is expressed as a velocity (meters per second). The change in velocity from moment to moment is expressed as an acceleration (meters per second per second). The position of an object at a particular time can be plotted on a graph. You can also graph the velocity and acceleration of the object versus time. A graph is a mathematical picture of the motion of an object. For this reason, it is important to understand how to interpret a graph of position, velocity, or acceleration versus time. In this activity you will plot a graph of motion in real-time, that is, as the motion is happening.

For more information see Cutnell & Johnson, Physics, 6th ed., Volume One, Chapter 2, Section 2.1 to 2.3.

Materials

Equipment Needed	Qty	Equipment Needed	Qty
PASPORT Motion Sensor (PS-2103)	1	Large Rod Base (ME-8735)	1
USB Link (PS-2100)	1	Motion Sensor Reflector Board	1
Support Rod, 45 cm (ME-8736)	1		

Setup

Computer Setup

1. Plug the *USB Link* into the computer's USB port.

2. Plug the *PASPORT Motion Sensor* into the USB Link. This will automatically launch the PASPORTAL window.

3. Choose the appropriate DataStudio configuration file entitled **01 Motion in One Dimension CF.ds** and proceed with the following instructions.

SAFETY REMINDER

- Follow the directions for using the equipment.

- You will be moving backwards for part of this activity. Clear the area behind you for at least 2 meters (about 6 feet).

**THINK SAFETY
ACT SAFELY
BE SAFE!**

Equipment Setup

1. Mount the Motion Sensor on a support rod so that it is aimed at your midsection when you are standing in front of the sensor. Make sure that you can move at least 2 meters away from the Motion Sensor.

2. Position the computer monitor so you can see the screen while you move away from the Motion Sensor.

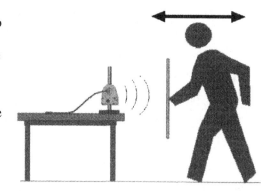

Record Data

Prepare to Record Your Position

(Hint: Read this section before you begin to take data.)

- When you are ready, stand in front of the Motion Sensor. You will click **Start** to begin.

- WARNING: You will be moving backward, so clear the area behind you.

- After you click **Start**, there is a three-second countdown before data recording begins. A 'pointer' on the vertical axis of the graph will move up and down as you move forward and backward relative to the sensor. Use the feedback from the 'pointer' to adjust your start position.

- The Motion Sensor will make a faint clicking noise.

- The digits display will show how closely you match the graph. The closer to 100, the better.

Record Your Position

1. Click **Start**.

2. Try to move so the plot of your motion matches the Position versus Time plot on the graph.

- Recording will stop automatically at 10 sec.

3. Repeat the data recording process a second and a third time. Try to improve the match.

Prepare to Record Your Velocity

- Now get ready to do the same procedure using a Velocity vs. Time graph.

- Stand in front of the Motion Sensor. You will click **Start** to begin.

- WARNING: You will be moving backward, so clear the area behind you.

- After you click **Start**, there is a three-second countdown before data recording begins.

- The Motion Sensor will make a faint clicking noise.

- The digits display will show how closely you match the graph. The closer to 100, the better.

Record Your Velocity

1. Click **Start**.

2. Try to move so the plot of your motion matches the Velocity versus Time plot on the graph.

- Recording will Stop automatically at 10 s.

3. Repeat the data recording process a second and a third time. Try to improve the match.

Analyze

Observations

1. How well did your motion graphs match the provided graphs?

2. What was the meaning of the part of the position plot where the slope was positive (upward)?

3. What was the meaning of the part of the velocity plot where the slope was positive (upward)?

4. Were certain parts of the plots easier to match than other parts? Why or why not?

Synthesize

Error Analysis

1. What are some sources of error in this exploration?

2. Why was your graph of position not an exact match of the provided plot?

Conclusions

1. What is the description of your motion for both graphs? (Example: "Constant speed for 2 seconds followed by no motion for 3 seconds, etc.")

2. Do your results support your hypothesis?

Applications

What are some physical phenomena where knowing your motion is important?

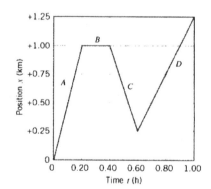

Extension Problem

In Cutnell & Johnson, Physics, 6th ed., Volume One, Chapter 2, problem 60, page 52.

A person who walks for exercise produces the position–time graph given with this problem.

(a) Without doing any calculations, decide which segments of the graph (A, B, C, or D) indicate positive, negative, and zero average velocities.

(b) Calculate the average velocity for each segment to verify your answers to part (a).

Activity 2: Position, Velocity, and Acceleration
(Motion Sensor)

Preface

- *If* you are using the PASCO electronic Workbook specifically designed for this activity, then do the following:
1. Connect the *USB Link* to the computer's USB port.
2. Connect the *Motion Sensor* to the USB Link. This will automatically launch the PASPORTAL window.
3. Choose the electronic Workbook entitled: **02 PVA WB.ds** and follow the directions in the Workbook.

Introduction

The purpose of this exploration is to study the relationships between position, velocity, and acceleration in linear motion and, in turn, how the relationships can be used to describe an object's motion.

Use the Motion Sensor to measure the motion of a cart as it moves away from the sensor.

Analyze the position and velocity of the cart on separate graphs. [The Analysis section has instructions for determining the acceleration of an object from its position vs. time and velocity vs. time graphs. Pay close attention to the appearance of the velocity graph (linear) and the position graph (quadratic).]

Learning Outcomes

You will be able to:

- Measure the position and velocity of an object that has uniform acceleration.

- Explore the relationships between position, velocity, and acceleration.

- Analyze the graphs of position and velocity vs. time for a cart's motion.

- Explain how the slope of the position graph relates to the velocity graph.

- Recognize the general shape of position and velocity graphs for motion involving constant acceleration.

Hypothesis

What is the shape of the position versus time graph for an object with uniform acceleration?

What is the shape of the velocity versus time graph for an object with uniform acceleration?

How could you determine an object's acceleration from a graph of velocity versus time?

Background

The equations of translational (or linear) kinematics can be used for any moving object, as long as the acceleration of the object is constant. The equations to the right can be used to solve problems when constant acceleration is present. Later in this exploration you will use these equations to answer some questions about the data you will record.

$$v = v_o + at$$

$$x = \frac{1}{2}(v_o + v)$$

$$x = v_o t + \frac{1}{2}at^2$$

$$v^2 = v_o + 2ax$$

For more information see Cutnell & Johnson, _Physics_, 6th ed. Volume One, Chapter 2, Section 2.4.

Materials

Equipment Needed	Qty	Equipment Needed	Qty
PASPORT Motion Sensor (PS-2103)	1	1.2 m PAScar Dynamics System (ME-6955)	1
USB Link (PS-2100)	1	Fan Accessory (ME-9491)	1

Setup

Computer Setup

1. Plug the _USB Link_ into the computer's USB port.

2. Plug the _PASPORT Motion Sensor_ into the USB Link. This will automatically launch the PASPORTAL window.

3. Choose the appropriate DataStudio configuration file entitled **02 PVA CF.ds** and proceed with the following instructions.

SAFETY REMINDER	
• Follow the directions for using the equipment. • Keep hands and objects away from the fan blade when it is turning.	**THINK SAFETY** **ACT SAFELY** **BE SAFE!**

Equipment Setup

1. Put batteries in two of the slots in the base of the Fan Accessory and put the metal rods in the other two slots. (Two batteries makes the fan run at low speed.)

2. Attach the Fan Accessory to the cart and place them on the track. Level the track so the cart will not roll on its own. (You could use a bubble level to make sure.)

3. Attach the Motion Sensor to the end of the track and make sure there is nothing to obstruct the signal going from the front of the Motion Sensor to the cart. Make sure the fan will push the cart away from the Motion Sensor.

4. Set the Motion Sensor Range Switch to 'Cart'.

Your setup may vary.

5. Hold the cart about 15 cm in front of the Motion Sensor.

6. Start the Fan Accessory (make sure the fan will push the cart away from the sensor.)

Record Data

Record the Motion

1. Click **Start** to begin recording data.

2. Release the cart so the fan can push the cart away from the Motion Sensor.

3. Click **Stop** when the cart has nearly reached the end of the track.

4. Turn off the Fan Accessory.

5. Click the **Scale to fit** button in your graph if necessary.

• If the plot is rough, try adjusting the tilt angle of the Motion Sensor and retake the set of data.

Analyze

Observations

1. What is the shape of the position versus time graph?

2. What is the shape of the velocity versus time graph?

3. How well did your predictions match the motion sensor generated plots?

How to Find the Acceleration from Position versus Time

1. Highlight a smooth part of the position graph.

2. Click the **Fit** menu and select **Quadratic Fit**.

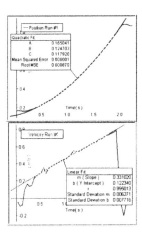

The "A" is the coefficient of the squared term in the curve fit formula. Record the coefficient "A".

How to Find the Acceleration from Velocity versus Time

1. Highlight a smooth part of the velocity graph.

2. Click the **Fit** menu and select **Linear Fit**.

The "m" value is the slope. Record the slope as the acceleration.

Compare the Acceleration Values

Multiply the value of "A" number by "2" to get the acceleration based on position versus time. Record the value of 2 x "A".

2 x A =

Compare this value to the slope ("m") of the Linear Fit for the velocity versus time graph. (This is the acceleration based on velocity versus time.)

Calculations

1. Using the acceleration from your position versus time data and the equations of motion from the Background section, calculate the time it would take for the cart to travel 50 meters.

2. Using the time you found in the previous problem calculate how fast the cart would be traveling.

Synthesize

Variables

What was the independent variable in this activity (what did you control)?

What variables did you measure?

Verification

Compute the percent difference of your two values for acceleration.

$$\% \text{ difference} = \left| \frac{\text{acceleration from position} - \text{acceleration from velocity}}{\text{acceleration from velocity}} \right| \times 100\%$$

»Acceleration from position versus time (2 x "A" value):

»Acceleration from velocity versus time (slope):

»Percent difference:

Error Analysis

What were the sources of error in this experiment?

Conclusions

1. In this exploration you used a graphical representation to show how position and velocity are related for an object with constant acceleration. What did you discover?

2. Do your results support your hypothesis?

Applications

What are some examples of where the equations of motion are used?

Extension Problem

In Cutnell & Johnson, Physics, 6th ed., Volume One, Chapter 2, concept question 85, page 54.

(a) The acceleration of a NASCAR racecar is zero. Does this necessarily mean that the velocity of the car is also zero? (b) If the speed of the car is constant as it goes around the track, is its average acceleration necessarily zero? Justify your answers.

Activity 3A: Projectile Motion Part 1 – Change Initial Speed
(Photogates and Time-of-Flight Pad)

Preface

- *If* you are using the PASCO electronic Workbook specifically designed for this activity, then do the following:
1. Connect the *USB Links* to the computer's USB port.
2. Connect the *Photogate Ports* to the USB Links. This will automatically launch the PASPORTAL window.
3. Choose the electronic Workbook entitled: **03A Projectile Motion Part 1 WB.ds** and follow the directions in the Workbook.

Introduction

The purpose of this exploration is to compare the time-of-flight of a projectile for different values of initial speed when a launcher is aimed horizontally.

Use Photogates and a Time-of-Flight Pad to measure the initial speed and the time-of-flight of a projectile.

Compare the time of flight for the projectile when it is launched horizontally at different initial speeds.

Learning Outcomes

You will be able to:

- Explore the basic concepts of projectile motion.

- Understand how gravitation and other forces are involved with projectile motion.

- Explain and justify the independence of horizontal and vertical directions of motion.

- Use experimental data along with basic kinematics equations to solve problems involving projectiles.

Hypothesis

What do you need to know in order to predict how long a ball will stay in the air? Explain your answer.

Does a change in the ball's initial speed change the time of flight? Explain your answer.

Background

The equations are used to study projectile motion. You will use these equations later in this exploration.

$$x - x_o = v_o \cos\theta_o t$$

Projectile motion is a kind of two-dimensional motion that occurs when the moving object (the projectile) experiences only the acceleration due to gravity, which acts in the vertical direction. The acceleration of the projectile has no horizontal component, the effects of air resistance being negligible. The vertical component of the acceleration equals the acceleration due to gravity.

$$y - y_o = v_o \sin\theta_o t - \frac{1}{2}gt^2$$

$$v_y = v_o \sin\theta_o - gt$$

$$v_y^2 = (v_o \sin\theta_o)^2 - 2g(y - y_o)$$

The vertical motion of a freely falling ball launched horizontally off a table of height 'd' is independent of any horizontal motion the ball may have. Thus the time for a ball to fall to the ground is independent of its horizontal speed. The distance 'd' a ball falls from rest as a function of the time of fall 't' is given by Equation 3.1 where 'g' is the acceleration in free fall.

$$d = \frac{1}{2}gt^2 \quad \text{Equation 3.1}$$

Thus the time for a ball to fall straight down a distance 'd' from rest to the ground is given by

$$t = \sqrt{\frac{2d}{g}} \quad \text{Equation 3.2}$$

If a ball launched horizontally with a non-zero initial speed takes the same amount of time to reach the ground as a ball that drops from rest from the same height, this equation also gives the time of flight for any ball launched horizontally regardless of the initial speed of the ball.

For more information see Cutnell & Johnson, Physics, 6th ed., Vol. One, Chapter 3, Section 3.3.

Materials

Equipment Needed	Qty	Equipment Needed	Qty
PASPORT Photogate Port (PS-2123)	2	Projectile Launcher (ME-6800)	1
Photogate Head (ME-9498)	2	Metric Measuring Tape, 30 m (SE-8712)	1
USB Link (PS-2100)	2	Extension Cord, 6 m (PI-8117)	1
Time of Flight Accessory Kit (ME-6810)	1	C-clamp, large	1
Photogate Mounting Bracket (ME-6821)	1		

Setup

Computer Setup

1. Plug the *USB Links* into the computer's USB port.

2. Plug the *PASPORT Photogate Ports* into the USB Links.

3. Plug the Photogate Heads into one of the Photogate Ports.

4. Plug the Phone Jack Extender Cable into the other Photogate Port. Plug the Time-of-Flight Pad into the Phone Jack Extender Cable. This will automatically launch the PASPORTAL window.

5.   Choose the DataStudio configuration file entitled **03A Projectile Motion Part 1 CF.ds** and proceed with the following instructions.

SAFETY REMINDER	THINK SAFETY
• Follow the directions for using the equipment. • Wear safety goggles when using the Projectile Launcher.	ACT SAFELY BE SAFE!

Equipment Setup

•   You can use the same setup for both Part 1 and Part 2 of Projectile Motion.

1.   Clamp the base of the projectile launcher to the edge of a sturdy table. Aim the launcher away from the table toward the center of an open area at least 3 meters away.

2.   Adjust the angle of the launcher to zero degrees so the plastic ball will be launched horizontally.

3.   Slide the photogate mounting bracket into the T-slot on the bottom side of the projectile launcher. Mount one Photogate to the bracket in the position closest to the end of the launcher. Mount the other Photogate to the bracket in the other position.

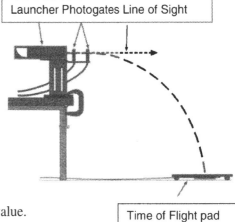

Launcher Photogates Line of Sight

Time of Flight pad

•   Note: Make sure the Photogates that are mounted on the front of the projectile launcher are separated by 10 centimeters (0.10 m). If the distance of separation is different, adjust the value. Click the Setup button and choose Change Value. Insert your value.

Record Data

Horizontal, Short Range

1.   Put the plastic ball into the projectile launcher. Cock the launcher to the short-range position.

2.   Test fire the ball to determine where to place the timing pad on the floor. Put the timing pad on the floor where the ball hits.

3.   Reload the ball into the projectile launcher, and cock the launcher to the short-range position. Click **Start** to begin recording data.

4.   Shoot the ball on the short-range position. After the ball hits the Time-of-Flight pad, click **Stop**. Record the initial speed and time of flight.

Horizontal, Middle Range

5.   Reload the ball into the launcher, but cock the launcher to the middle-range position. Test fire the ball to determine the new location to put the Time-of-Flight pad. Move the pad.

6.   Reload the ball into the launcher and put the launcher in the middle-range position.

7.   When you are ready, click **Start**.

8. Shoot the ball with the launcher in the middle-range position. After the ball hits the Time-of-Flight pad, click **Stop**. Record the initial speed and time of flight.

Horizontal, Long Range

9. Reload the ball into the launcher, but cock the launcher to the long-range position. Test fire the ball to determine the new location to put the Time-of-Flight pad. Move the pad.

10. Reload the ball into the launcher and put the launcher in the long-range position.

11. When you are ready, click **Start**.

12. Shoot the ball with the launcher in the long-range position. After the ball hits the Time-of-Flight pad, click **Stop**. Record the initial speed and time of flight.

Analyze

Observations

How do the values for the time of flight for the short, middle, and long-range distances compare when the ball was launched horizontally?

Synthesize

Variables

What was the independent variable in this exploration (what did you change from one run to the next)?

Which variables did you measure?

What was the response when you changed the independent variable?

Error Analysis

What were the sources of error in this experiment?

Conclusions

1. How can you predict how long a ball will stay in the air? Does a change in its initial speed change the "time of flight"? If so, how?

2. Do your results support your hypothesis?

Applications

Give some real world examples of projectile motion?

Extension Problem

In Cutnell & Johnson, Physics, 6th ed., Volume One, Chapter 3, additional problem 62, page 78.

A bullet is fired from a rifle that is held 1.6 m above the ground in a horizontal position. The initial speed of the bullet is 1100 m/s. Find (a) the time it takes for the bullet to strike the ground and (b) the horizontal distance traveled by the bullet.

Activity 3B: Projectile Motion Part 2 – Change Launch Angle
(Photogates and Time-of-Flight Pad)

Preface

Introduction

The purpose of this exploration is to compare the time-of-flight of a projectile for different initial speeds when a launcher is aimed at an angle above horizontal.

Use two Photogates and a Time-of-Flight Pad to measure the initial speed and the time-of-flight of a projectile.

Predict the range of a projectile based on the initial speed, vertical height, and launch angle.

Learning Outcomes

You will be able to:

- Explore the basic concepts of projectile motion.

- Understand how gravity and other forces are involved with projectile motion.

- Explain and justify the independence of horizontal and vertical directions of motion.

- Use experimental data along with basic kinematics equations to solve problems involving projectiles.

Hypothesis

What do you need to know in order to predict how long a ball will stay in the air? Explain.

Does a change in the ball's initial speed change the time of flight or the range? Explain.

Does a change in angle affect the time of flight or the range? If so, how?

Background

The equations to the right are used to study projectile motion. You will use these equations later in this exploration.

$$x - x_o = v_o \cos\theta_o t$$

$$y - y_o = v_o \sin\theta_o t - \frac{1}{2}gt^2$$

$$v_y = v_o \sin\theta_o - gt$$

$$v_y^2 = (v_o \sin\theta_o)^2 - 2g(y - y_o)$$

Projectile motion is a kind of two-dimensional motion that occurs when the moving object (the projectile) experiences only the acceleration due to gravity, which acts in the vertical direction. The acceleration of the projectile has no horizontal component ($a_x = 0$), the effects of air resistance being negligible. The vertical component of the acceleration is a_y, and it equals the acceleration due to gravity.

The vertical motion of a freely falling ball launched horizontally off a table of height 'd' is independent of any horizontal motion the ball may have. Thus the time for a ball to fall to the ground is independent of its horizontal speed. The distance 'd' a ball falls from rest as a function of the time of fall 't' is given by Equation 3.1 where 'g' is the acceleration due to gravity in free fall.

$$d = \frac{1}{2}gt^2 \quad \text{Equation 3.1}$$

Thus Equation 3.2 gives the time for a ball to fall straight down a distance 'd' from rest to the ground.

$$t = \sqrt{\frac{2d}{g}} \quad \text{Equation 3.2}$$

If a ball launched horizontally with a non-zero initial speed takes the same amount of time to reach the ground as a ball that drops from rest from the same height, this equation also gives the time of flight for any ball launched horizontally regardless of the initial speed of the ball.

The equations to the right are used to calculate range. You will use these equations later in this exploration.

$$v_{ox} = v_o \cos\theta$$

$$v_{oy} = v_o \sin\theta$$

For more information see Cutnell & Johnson, Physics, 6th ed., Volume One, Chapter 3, Section 3.3.

$$x = R = 2t \times v_{ox}$$

Materials

Equipment Needed	Qty	Equipment Needed	Qty
PASPORT Photogate Port (PS-2123)	2	Projectile Launcher (ME-6800)	1
Photogate Head (ME-9498)	2	Metric Measuring Tape, 30 m (SE-8712)	1
USB Link (PS-2100)	2	Extension Cord, 6 m (PI-8117)	1
Time of Flight Accessory Kit (ME-6810)	1	C-clamp, large	1
Photogate Mounting Bracket (ME-6821)	1		

Setup

Computer Setup

1.   Plug the *USB Links* into the computer's USB port.

2.   Plug the *PASPORT Photogate Ports* into the USB Links.

3. Plug the Photogate Heads into one of the Photogate Ports.

4. Plug the Phone Jack Extender Cable into the other Photogate Port. Plug the Time-of-Flight Pad into the Phone Jack Extender Cable. This will automatically launch the PASPORTAL window.

5. Choose the DataStudio configuration file entitled **03B Projectile Motion Part 2 CF.ds** and proceed with the following instructions.

SAFETY REMINDER	THINK SAFETY
• Follow the directions for using the equipment. • Wear safety goggles when using the Projectile Launcher.	**ACT SAFELY** **BE SAFE!**

Equipment Setup

1. Clamp the base of the projectile launcher to the edge of a sturdy table. Aim the launcher away from the table toward the center of an open area at least 3 meters away.

2. Adjust the angle of the launcher to zero degrees so the plastic ball will be launched horizontally.

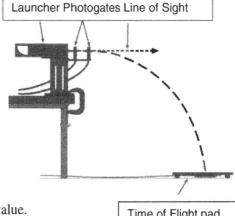

3. Slide the photogate mounting bracket into the T-slot on the bottom side of the projectile launcher. Mount one Photogate to the bracket in the position closest to the end of the launcher. Mount the other Photogate to the bracket in the other position.

•   Note: Make sure the Photogates that are mounted on the front of the projectile launcher are seperated by 10 centimeters (0.10 m). If the distance of separation is different, adjust the value. Click the Setup button and choose Change Value. Insert your value.

Record Data

Horizontal, Short Range

1. Put the plastic ball into the projectile launcher. Cock the launcher to the short-range position.

2. Test fire the ball to determine where to place the timing pad on the floor. Put the timing pad on the floor where the ball hits.

3. Reload the ball into the projectile launcher, and cock the launcher to the short-range position. Click **Start** to begin recording data.

4. Shoot the ball on the short-range position. After the ball hits the Time-of-Flight pad, click **Stop**. Record the initial speed and time of flight.

30 degrees, Short Range

1. Adjust the angle of the launcher to 30° above horizontal.

2. Reload the launcher and set it to the short-range position. Test fire the ball to determine where to place the timing pad on the floor. Put the timing pad on the floor where the ball hits.

3. Reload the launcher and set it to the short-range position. Click **Start** and shoot the ball.

4. After the ball hits the timing pad, click **Stop**.

5. Record the Initial Velocity. You will use the value to predict horizontal range.

Initial Velocity =

30 degrees, Middle Range

1. Leave the angle of the launcher at 30° above horizontal.

2. Reload the launcher and set it to the middle-range position. Test fire the ball to determine where it lands and move the timing pad to that spot.

3. Reload the launcher and set it to the middle-range position. Click **Start** and shoot the ball.

4. After the ball hits the timing pad, click **Stop**. Record the initial speed and time of flight.

30 degrees, Long Range

1. Leave the angle of the launcher at 30° above horizontal.

2. Reload the launcher and set it to the long-range position. Test fire the ball to determine where it lands and move the timing pad to that spot.

3. Reload the launcher and set it to the long-range position. Click **Start** and shoot the ball.

4. After the ball hits the timing pad, click **Stop**. Record the initial speed and time of flight.

Predict the Range and Test the Prediction

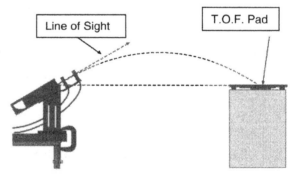

1. Based on the Initial Velocity and the launch angle (30°), calculate the initial vertical velocity and the initial horizontal velocity.

2. Based on the initial vertical velocity, find the time, 't', it takes for the projectile to reach its maximum height. Multiply the time by '2' to get the total time in flight.

3. Calculate the range based on the initial horizontal velocity and the total time in flight.

Initial Velocity		Time (max. height)	
Initial Horizontal Velocity		2 x Time	
Initial Vertical Velocity		Predicted Range	

4. Place the Time-of-Flight Pad at the horizontal distance you calculated for the range. Place the pad at the same height as the end of the projectile launcher.

5. Reload the launcher and set it to the short-range position. Shoot the ball and observe how close it lands to the spot you predicted.

Analyze

Observations

1. How do the values for the time of flight for the short-range, horizontal compare to the short-range, 30° angle distance?

2. How do the values for the time of flight for the short, middle, and long-range distances compare when the ball was launched at 30° above the horizon?

3. Why would time of flight depend on the angle of the launch?

4. When you tested the range, how close did the ball land to the spot you predicted?

Synthesize

Variables

1. What was the independent variable in this exploration (what did you change from one run to the next)?

2. Which variables did you measure?

3. What was the response when you changed the independent variable?

Error Analysis

What were the sources of error in your experiment?

Conclusions

1. How can you predict how long a ball will stay in the air? Does a change in its angle change the "time of flight"? If so, how?

2. Do your results support your hypothesis?

Applications

Give some real world examples of projectile motion?

Extension Problem

In Cutnell & Johnson, Physics, 6th ed., Volume One, Chapter 3, problem 27, page 76.

A motorcycle daredevil is attempting to jump across as many buses as possible. The takeoff ramp makes an angle of 18.0 degrees above the horizontal, and the landing ramp is identical to the takeoff ramp. The buses are parked side by side, and each bus is 2.74 m wide. The cyclist leaves the ramp with a speed of 33.5 m/s. What is the maximum number of buses over which the cyclist can jump?

4A: Newton's Second Law Part 1 – Constant Mass
(Motion Sensor)

Preface

> • *If* you are using the PASCO electronic Workbook specifically designed for this activity, then do the following:
> 1. Connect the *USB Link* to the computer's USB port.
> 2. Connect the *Motion Sensor* to the USB Link. This will automatically launch the PASPORTAL window.
> 3. Choose the electronic Workbook entitled: **04A Newton's 2nd Part 1 WB.ds** and follow the directions in the Workbook.

Introduction

One of the important equation in physics is $a = F/m$, also known as Newton's Second Law of motion. Newton's Second Law describes the behavior of everything that changes its motion due to a net force -- from the trajectory of a baseball to the motion of a planet.

The purpose of this exploration is to find out what happens to an object's acceleration when the net force applied to the object increases and the mass of the system is constant.

Use a Motion Sensor to measure the motion of an object that is accelerated by a net force. Determine what happens to the acceleration of the cart when the net force is increased and the mass stays constant.

Note: In Newton's Second Law Part 2 you will measure the acceleration of an object when the net force is kept constant but the mass is increased.

Learning Outcomes

You will be able to:

• Interpret data from a graph of velocity versus time to find acceleration.

• Measure the acceleration of an object with a constant mass when the net force is changed.

• Calculate theoretical values for acceleration based on net force and mass and compare the theoretical values with the measured values.

• Describe the relationship between the acceleration of an object and the net force applied to the object.

Hypothesis

What happens to an object when you apply a net force to it?

What happens to the motion when you change the magnitude of that net force?

Background

Newton's First Law states that if no net force acts on an object, then the velocity of the object remains unchanged. The Second Law deals with what happens when a net force does act.

As long as a net force acts, the velocity of an object changes - in other words, it accelerates. If more force is applied, the greater force produces a greater acceleration. Twice the force produces twice the acceleration.

$$a \propto F_{net}$$

Often, several forces act on an object simultaneously. In such cases, it is the net force, or the vector sum of all the forces acting, that is important. Newton's second law states that the acceleration is proportional to the net force acting on the object.

$$a \propto \frac{1}{m}$$

Newton's Second Law also states that the acceleration is inversely proportional to the mass.

$$a = \frac{F_{net}}{m}$$

For more information see Cutnell & Johnson, Physics, 6th ed., Vol. One, Chapter 4, Section 4.3.

Materials

Equipment Needed	Qty	Equipment Needed	Qty
PASPORT Motion Sensor (PS-2103)	1	Balance (SE-8707)	1
USB Link (PS-2100)	1	String (SE-8050)	1 m
Mass and Hanger Set (ME-9348)	1	Super Pulley with Clamp (ME-9448)	1
1.2 m PAScar Dynamics System (ME-6955)	1		

Setup

Computer Setup

1. Plug the *USB Link* into the computer's USB port.

2. Plug the *PASPORT Motion Sensor* into the USB Link. This will automatically launch the PASPORTAL window.

3. Choose the DataStudio configuration file entitled **04A Newton's 2nd Part 1 CF.ds** and proceed with the following instructions.

SAFETY REMINDER	THINK SAFETY
• Follow the directions for using the equipment.	ACT SAFELY BE SAFE!

Equipment Setup

• You can use the same equipment setup for both Part 1 and Part 2 of Newton's Second Law.

1. The track has a fixed end stop at one end. Place the Motion Sensor on the track in front of the fixed end stop.

2. Attach the Pulley to the other end of the track. If you have a piece of wood you can place this in front of the pulley to stop the cart.

3. Level the track. Put the cart on the track. If the cart moves one way or the other, use the leveling screw under the fixed end stop to level the track.

4. Attach a piece of string about 1.2 m long to one end of the cart.

- Note: It is very important that the track is level to get the best results.

5. Add a 20-g mass to the mass hanger and weigh the hanger plus the mass. Record the mass.

Run #1: Total mass of the hanger and 20-g mass (mH):

6. Put two 20-g masses into the accessory tray on top of the cart.

7. Weigh the cart plus the two 20-g masses and record the total mass of the cart and masses.

Run #1: Total mass of the cart and two 20-g masses (mC):

8. Attach the mass hanger to the other end of the string, and put the string in the pulley's groove.

9. Adjust the pulley up or down so the string is parallel to the track as shown in the diagram.

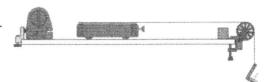

10. Hold the cart in front of the Motion Sensor but no closer than 15 cm from the sensor.

Record Data

Constant Mass

- You will take three runs of data. For Run #1, use the arrangement described above (one 20-g mass on the mass hanger and two 20-g masses on top of the cart).

- For the second and third runs, transfer mass from the tray of the cart to the mass hanger on the string.

- For Run #2, transfer one 20-g mass from the tray of the cart to the mass hanger before you record data. The total mass of the system is constant, but the net force is increased.

- For Run #3, transfer a second 20-g mass from the tray of the cart to the mass hanger before you record data. As in Run #2, the total mass of the system is constant, but the net force is increased.

- Note: Catch the cart before it hits the pulley or you could damage the pulley.

1. Click **Start**. Release the cart. Stop the cart before it hits the pulley and click **Stop**.

- The analyze section describes how to determine the acceleration.

2. For Run #2, move a 20-g mass from the cart to the hanger. Record the total mass of the hanger. Record data as before.

3. For Run #3, move the remaining 20-g mass from the cart to the hanger. Record the total mass of the hanger. Record data as before.

Run #2: Total of hanger plus masses:

Run #3: Total of hanger plus masses:

Analyze

Examine the Graph

1. Click **Scale to Fit** to rescale the graph if needed. Use the cursor to highlight a smooth part of the graph.

2. Click the **Fit** menu and choose **Linear Fit**.

*   The slope "m" of the best-fit line is the acceleration.

3. After recording the slope for Run #1, use the **Data** menu to select the next run.

*   This example shows the highlighted area and the **Linear Fit** information box.

4. Record the value of "m" (slope) as the acceleration for Run #2.

5. After recording the slope, use the **Data** menu to select the next run.

6. Repeat the process for Run #3.

Run	Acceleration (m/s/s)
#1	
#2	
#3	

Observations

1. What did you observe about the slope of the Linear Fit as the net force increased but the total mass was kept constant?

2. Why did the slope change for each run?

Calculations: Constant Mass, Increasing Net Force

Calculate the theoretical acceleration when mass is constant and net force is changed.

The acceleration is the ratio of the net force divided by the total mass.

For runs #1, #2, and #3, the total mass of the system (mass of cart plus mass of hanger) is constant and the net force (mass of hanger x 9.8) increases.

$$a = \frac{m_H g}{m_C + m_H}$$

Assuming no friction, the net force is the weight of the hanger (mass x 9.8 N/kg).

Run	Mass, hanger (kg)	Net Force (N)	Acc., theory (m/s^2)	Acc., exp (m/s^2)	% difference
#1					
#2					
#3					

Total mass of cart and two 20-g masses (mC):

Total mass of system (mC + mH) where mH is the mass of the hanger for Run #1:

Synthesize

Variables

1. What variable(s) did you keep the same from one run to the next? What variable(s) did you control?

2. What variable changed as a result and what was the response?

Error Analysis

What were the sources of error in your experiment?

Conclusions

What happens to an object's acceleration if the net force applied to the object is increased but the object's mass remains constant?

Do your results support your hypothesis?

Applications

Based on Newton's Second Law, predict what will happen to the Space Shuttle's acceleration after lift off as it burns its fuel.

Extension Problem

In Cutnell & Johnson, Physics, 6th ed., Volume One, Chapter 4, problem 3, page 116.

In the amusement park ride known as Magic Mountain Superman, powerful magnets accelerate a car and its riders from rest to 45 m/s (about 100 mi/h) in a time of 7.0 s. The mass of the car and riders is 5,500 kg. Find the average net force exerted on the car and riders by the magnets.

Activity 4B: Newton's Second Law Part 2 – Constant Net Force
(Motion Sensor)

Preface

- *If* you are using the PASCO electronic Workbook specifically designed for this activity, then do the following:
1. Connect the *USB Link* to the computer's USB port.
2. Connect the *Motion Sensor* to the USB Link. This will automatically launch the PASPORTAL window.
3. Choose the electronic Workbook entitled: **04B Newton's 2nd Part 2 WB.ds** and follow the directions in the Workbook.

Introduction

One of the important equation in physics is $a = F/m$, also known as Newton's Second Law of motion. Newton's Second Law describes the behavior of everything that changes its motion due to a net force -- from the trajectory of a baseball to the motion of a planet.

The purpose of this exploration is to find out what happens to an object's acceleration when the net force applied to the object remains constant and the mass of the system is increased.

Use a Motion Sensor to measure the motion of an object that is accelerated by a net force. Determine what happens to the acceleration of the cart when the net force is constant and the mass increases.

Note: In Newton's Second Law Part 1 you measured the acceleration of an object when the net force is increased but the mass of the system was constant.

Learning Outcomes

You will be able to:

- Interpret data from a graph of velocity versus time to find acceleration.

- Measure the acceleration of an object when a constant net force is applied and the mass is increased.

- Calculate theoretical values for acceleration based on net force and mass and compare the theoretical values with the measured values.

- Describe the relationship between the acceleration of an object, the net force applied to the object, and the mass of the object.

Hypothesis

What happens to an object when you apply a net force to it?

What happens to the motion of an object if you change the mass of the object but keep the net force constant?

Background

Newton's First Law states that if no net force acts on an object, then the velocity of the object remains unchanged. The Second Law deals with what happens when a net force does act.

$$a \propto F_{net}$$

$$a \propto \frac{1}{m}$$

$$a = \frac{F_{net}}{m}$$

As long as a net force acts, the velocity of an object changes — in other words, it accelerates. If more force is applied, the greater force produces a greater acceleration. Twice the force produces twice the acceleration.

In Newton's second law, the net force is only one of two factors that determine the acceleration. The other is the inertia (or mass) of the object. Newton's Second Law states that for a given net force, the magnitude of the acceleration is inversely proportional to the mass. Twice the mass means one-half the acceleration, if the same net force acts on the object.

The second law shows how the acceleration depends on both the net force and the mass.

For more information see Cutnell & Johnson, Physics, 6th ed., Vol. One, Chapter 4, Section 4.3.

Materials

Equipment Needed	Qty	Equipment Needed	Qty
PASPORT Motion Sensor (PS-2103)	1	Balance (SE-8723)	1
USB Link (PS-2100)	1	String (SE-8050)	1 m
Mass and Hanger Set (ME-9348)	1	Super Pulley with Clamp (ME-9448)	1
1.2 m PAScar Dynamics System (ME-6955)	1		

Setup

Computer Setup

1. Plug the *USB Link* into the computer's USB port.

2. Plug the *PASPORT Motion Sensor* into the USB Link. This will automatically launch the PASPORTAL window.

3. Choose the DataStudio configuration file entitled **04B Newton's 2nd Part 2 CF.ds** and proceed with the following instructions.

SAFETY REMINDER	THINK SAFETY
• Follow the directions for using the equipment.	ACT SAFELY BE SAFE!

Equipment Setup

1. The track has a fixed end stop at one end. Place the Motion Sensor on the track in front of the fixed end stop.

2. Attach the Pulley to the other end of the track. If you have a piece of wood you can place this in front of the pulley to stop the cart.

3. Level the track. Put the cart on the track. If the cart moves one way or the other, use the leveling screw under the fixed end stop to level the track.

4. Attach a piece of string about 1.2 m long to one end of the cart.

- Note: It is very important that the track is level to get the best results.

5. Add a 20-g mass to the mass hangar and weigh the hanger plus the mass. Record the mass.

Run #1: Total mass of the hanger and the 20-g mass (mH):

6. Weigh the cart and record the total mass of the cart.

Run #1: Total mass of the cart (mC):

7. Attach the mass hanger to the other end of the string, and put the string in the pulley's groove.

8. Adjust the pulley up or down so the string is parallel to the track as shown in the diagram.

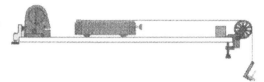

9. Hold the cart in front of the Motion Sensor but no closer than 15 cm from the sensor.

Record Data

Constant Net Force

- You will take three runs of data. For Run #1, use the arrangement described above (one 20-g mass on the mass hanger).

- For the second and third runs, add mass to the tray on the cart.

- For Run #2, add one 250-g mass bar to the tray of the cart before you record data. The total mass of the system is increased, but the net force is the same.

- For Run #3, add a second 250-g mass bar to the tray of the cart before you record data. As in Run #2, the total mass of the system is increased, but the net force is kept constant.

- Note: Catch the cart before it hits the pulley or you could damage the pulley.

1. Click **Start**. Release the cart. Stop the cart before it hits the pulley and click **Stop**.

- The analyze section describes how to determine the acceleration.

2. For Run #2, add a 250-g mass to the cart. Record the total mass of the cart plus the extra mass. Record data as before.

3. For Run #3, add a second 250-g mass to the cart. Record the total mass of the cart plus the two extra masses. Record data as before.

Run #2: Total of cart plus one extra mass:

Run #3: Total of cart plus two extra masses:

Analyze

Examine the Graph

1. Click the **Scale to Fit** button if needed. Highlight the region of the velocity plot that shows the motion of the cart.

2. Click the **Fit** menu and select **Linear Fit**.

- The slope "m" of the best-fit line is the acceleration.

3.   After recording the slope (acceleration) for Run #1, use the **Data** menu to select the next run.

•   This example shows the highlighted area and the Linear Fit information box.

4.   Select Run #2. Repeat the process to find the acceleration. After recording the value, use the **Data** menu to select Run #3. Repeat the process.

Linear Fit	
m (Slope)	0.12702
b (Y Intercept)	-0.12286
r	0.99335
Standard Deviation m	0.00658
Standard Deviation b	0.00801

Run	Acceleration (m/s/s)
#1	
#2	
#3	

Observations

What did you observe about the slope of the Linear Fit as the net force remained the same but the total mass was increased?

Calculations: Increasing Mass, Constant Net Force

Calculate the theoretical acceleration when mass increases and net force stays the same.

The acceleration is the ratio of the net force divided by the total mass.

$$a = \frac{m_H g}{m_C + m_H}$$

For runs #1, #2, and #3, the total mass of the system (mass of cart plus mass of hanger) increases and the net force (mass of hanger x 9.8) remains constant.

Assuming no friction, the net force is the weight of the hanger (mass of hanger x 9.8 N/kg).

Find the percent difference between the theoretical acceleration and the experimental acceleration.

Run	Mass, cart (kg)	Total mass (kg)	Acc., theory (m/s^2)	Acc., exp (m/s^2)	% difference
#1					
#2					
#3					

Total mass of hanger and one 20-g mass (mH):

Net force (mass of hanger x 9.8 N/kg):

Synthesize

Variables

1. What variable(s) did you keep the same from one run to the next? What variable(s) did you control?

2. What variable changed as a result and what was the response?

Error Analysis

What were the sources of error in your experiment?

Conclusions

What happens to an object's acceleration if the net force applied to the object is constant but the total mass increases?

Do your results support your hypothesis?

Applications

Based on Newton's Second Law, predict what will happen to a truck's acceleration if it gains mass during a rainstorm but the net force on the truck remains constant.

Extension Problem

An unloaded sled with a mass of 12 kg has an acceleration of 2.3 m/s/s when it is pushed on a horizontal surface with a net force of 27.6 N. What is its acceleration when it is fully loaded to 25 kg and the net force is the same?

Activity 5A: Newton's Third Law Part 1 – Collisions
(Force Sensors)

Preface

Introduction

The purpose of this exploration is to determine the forces exerted on two objects in a collision. You will measure and compare the forces on each of the two bodies.

Use Force Sensors attached to carts to measure the force during a head-on collision between the two carts.

Note: In Newton's 3rd Law Part 2 you will measure the forces of a tug-of-war between the two Force Sensors.

Learning Outcomes

You will be able to:

- Compare the force exerted on one cart during a collision to the force exerted on the second cart during the collision.

- Use a graph of force versus time for each cart to determine the force exerted on each cart during a collision.

- Demonstrate a quantitative understanding of Newton's Third Law by applying it to a problem related to collisions.

Hypothesis

How does the force exerted on one cart during a collision compare to the force exerted on the second cart in the collision?

If the carts have different masses, how will the force exerted by the heavier cart on the lighter cart compare to the force exerted by the lighter cart on the heavier cart?

Background

Whenever one body exerts a force on a second body, the second body exerts an oppositely directed force of equal magnitude on the first body.

The third law is often called the "action-reaction" law, for it is sometimes quoted as follows: "For every action (force) there is an equal, but opposite, reaction."

The diagram illustrates how the third law applies to an astronaut who is drifting just outside a spacecraft and who pushes on the spacecraft with a force P. According to the third law, the spacecraft pushes back on the astronaut with a force -P that is equal in magnitude but opposite in direction.

For more information see Cutnell & Johnson, Physics, 6th ed., Vol. One, Chapter 4, Section 4.5.

Materials

Equipment Needed	Qty	Equipment Needed	Qty
PASPORT Force Sensor (PS-2104)	2	1.2 m PAScar Dynamics System (ME-6955)	1
USB Link (PS-2100)	2	Compact Cart Mass (ME-6755)	2

Setup

Computer Setup

1. Plug the *USB Links* into the computer's USB port.

2. Plug the *PASPORT Force Sensors* into the USB Links. This will automatically launch the PASPORTAL window.

3. Choose the DataStudio configuration file entitled **05A Newton's 3rd Part 1 CF.ds** and proceed with the following instructions.

SAFETY REMINDER	THINK SAFETY
• Follow the directions for using the equipment.	ACT SAFELY BE SAFE!

Equipment Setup

1. Place the track on a horizontal surface. Level the track by placing the GOcar on the track. If the cart rolls one way or the other, use the leveling screw at one end of the  track to raise or lower that end until the track is level and the cart does not roll one way or the other.

2. Use the thumbscrew that comes with the Force Sensor to mount the sensor onto the accessory tray of each cart.

3. Remove the hooks from the Force Sensors.

4. Attach the cylindrical rubber bumper that comes with the Force Sensor to the front of each sensor.

• Note: It is very important that the track is level to get the best results.

5. The idea is to push the two carts together so they will collide. The Force Sensors will measure the forces exerted on each cart during the collision.

6. After you measure the forces during a collision for carts of equal mass, add 500 g of mass to one cart and then repeat the collision.

• Note: Do not push the carts together too fast or they may slip off the track during the collision.

7. The graph of force versus time is set up so that the measurements from the two sensors will be plotted on opposite sides of the X-axis. (One sensor is "Force, push right" and the other sensor is "Force, push left".)

• It is important to Zero both of the Force Sensors prior to each data run. When the equipment is ready, press the Zero button on top of each sensor.

Record Data

Collision: Equal Mass

1. Zero the sensors. Click **Start**. Push the carts and let them roll together. Click **Stop** after the collision.

• The analyze section shows how to interpret your force data.

2. For Run #2, add two 250-gram mass bars to one of the carts.

3. Zero the sensors. Click **Start**. Repeat the data recording procedure again, but allow the lighter cart to remain stationary and push only the heavier cart. Click **Stop** after the collision.

Analyze

Examine the Graph

1. Click the **Scale to Fit** button if needed.

2. Use the **Zoom Select** tool to highlight both peaks.

• The Legend box shows the minimum (Min.) and maximum (Max.) for both sensors.

3. Record the Min. value for "Force, push left" and the Max. value for "Force, push right" for Run #1.

• In this example they are -40.357 N and 40.592 N.

4. Use the Data menu to select Run #2 for both sensors. Repeat the procedure to find the values for "Force, push right" and for "Force, push left". Record the forces for each sensor for Run #2.

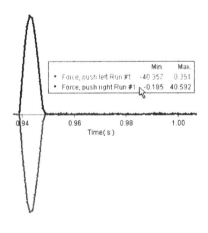

Run	Force, push left (N)	Force, push right (N)
#1		
#2		

Observations

1. What did you observe about the shape of the graphs of force versus time for each sensor?

2. What happened to the shape of the graphs of force from Run #2 when the heavier cart collided with the lighter cart?

Comparisons

1. Which cart experiences more force in a collision when both carts are moving and have equal masses?

2. Which cart experiences more force in a collision when a heavier cart that is moving collides with a lighter cart that is not moving?

3. What effect does the mass of a cart have on the force it experiences?

Synthesize

Variables

1. What was the independent variable in this activity (what did you change)?

2. What variables did you measure?

3. How did the measured variable change?

Conclusions

1. How does the force exerted on one cart during a collision compare to the force exerted on the second cart in the collision?

2. If the carts have different masses, how will the force exerted by the heavier cart on the lighter cart compare to the force exerted by the lighter cart on the heavier cart?

3. Do your results support your hypothesis?

Applications

What are examples of events where Newton's Third Law applies?

Extension Problem

In Cutnell & Johnson, Physics, 6th ed. Volume One, Chapter 4, problem 14, pages 116-7.

Airplane flight recorders must be able to survive catastrophic crashes. Therefore, they are typically encased in crash-resistant steel or titanium boxes that are subjected to rigorous testing. One of the tests is an impact shock test, in which the box must be able to survive being thrown at high speeds against a barrier. A 41-kg box is thrown at a speed of 220 m/s and is brought to a halt in a collision that lasts for a time of 6.5 ms. What is the magnitude of the average net force that acts on the box during the collision?

Activity 5B: Newton's Third Law Part 2 – Tug-of-War
(Force Sensors)

Preface

- *If* you are using the PASCO electronic Workbook specifically designed for this activity, then do the following:
1. Connect the *USB Links* to the computer's USB port.
2. Connect the *Force Sensors* to the USB Links. This will automatically launch the PASPORTAL window.
3. Choose the electronic Workbook entitled: **05B Newton's 3rd Part 2 WB.ds** and follow the directions in the Workbook.

Introduction

The purpose of this exploration is to determine the forces exerted on two objects in a tug-of-war. You will measure and compare the forces on each of the two bodies.

For this activity, use Force Sensors attached to carts to measure the force on each cart during a tug-of-war between the two carts.

Note: In Newton's 3rd Law Part 1 you measured the forces of a collision between the two Force Sensors.

Learning Outcomes

You will be able to:

- Compare the force exerted on one cart during a tug-of-war to the force exerted on the second cart during the tug-of-war.

- Use a graph of force versus time for each cart to determine the force exerted on each cart during a tug-of-war.

- Describe Newton's Third Law of Motion in conceptual terms.

- Demonstrate a quantitative understanding of Newton's Third Law by applying it to a problem related to a tug-of-war.

Hypothesis

How does the force on one end of the rope in a traditional tug-of-war compare to the force on the other end of the rope (assuming that the rope is in static equilibrium)?

How do the forces measured by two force sensors pulling away from each other compare? Are they equal?

Background

Whenever one body exerts a force on a second body, the second body exerts an oppositely directed force of equal magnitude on the first body.

The third law is often called the "action-reaction" law, for it is sometimes quoted as follows: "For every action (force) there is an equal, but opposite, reaction."

In a traditional tug-of-war, the winning team is usually the one that has the best traction against the ground. The forces exerted by each team pulling on the rope are equal in magnitude and opposite in direction.

For more information see Cutnell & Johnson, Physics, 6th ed., Vol. One, Chapter 4, Section 4.5.

Materials

Equipment Needed	Qty	Equipment Needed	Qty
PASPORT Force Sensor (PS-2104)	2	1.2 m PAScar Dynamics System (ME-6955)	1
USB Link (PS-2100)	2	String (SE-8050)	0.2 m

Setup

Computer Setup

1. Plug the *USB Links* into the computer's USB port.

2. Plug the *PASPORT Force Sensors* into the USB Links. This will automatically launch the PASPORTAL window.

3. Choose the DataStudio configuration file entitled **05B Newton's 3rd Part 2 CF.ds** and proceed with the following instructions.

SAFETY REMINDER	
• Follow the directions for using the equipment.	**THINK SAFETY ACT SAFELY BE SAFE!**

Equipment Setup

1. Place the track on a horizontal surface. Level the track by placing the GOcar on the track. If the cart rolls one way or the other, use the leveling screw at one end of the track to raise or lower that end until the track is level and the cart does not roll one way or the other.

2. Use the thumbscrew that comes with the Force Sensor to mount the sensor onto the accessory tray of each cart.

3. Put a hook on the front of each sensor. Place the carts on the track so the hooks face one another.

4. Tie a loop of string around the hooks on the Force sensors. The string should be about 20 cm.

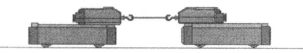

- Note: It is very important that the track is level to get the best results.

- It is important to Zero the Force Sensors prior to each data run.

- The idea is to pull the two carts apart as if in a tug-of-war. The Force Sensors will measure the forces during the pulling.

- The graph of force versus time is set up so that the measurements from the two sensors will be plotted on opposite sides of the X-axis. (One sensor is "Force, pull right" and the other sensor is "Force, pull left".)

Record Data

Tug-of-War

1. Zero both sensors and click **Start**. Slowly pull the carts apart. (Do not exceed 50 N or you could damage the sensors.)

2. Slowly release the carts and click **Stop**.

- The next section shows ways to analyze the force versus time data.

Analyze

Examine the Graph

1. Click **Scale to Fit** if needed.

2. Click "Force, pull right Run #1" in the graph's Legend box to select the data for that sensor.

3. Click **Smart Tool**. Move the **Smart Tool** cursor to a point on the "Force, pull right" curve. The cursor shows the time (x-coordinate) and the force (y-coordinate).

4. Click "Force, pull left Run #1" in the Legend box to select the data for the other sensor.

5. Click **Smart Tool** again. Move the **Smart Tool** cursor to the point on the "Force, pull left" curve that has the same time (x-coordinate) as the point on the "Force, pull right" curve.

6. Use the "left-right" arrow keys on the keyboard to move both Smart Tools along the force curves.

7. Observe how the force values compare. Try to find a point where the force values are equal but opposite. (In the example, the "Force, pull right Run #1" is 4.511 N and the "Force, pull left Run #1" is –4.440 N.)

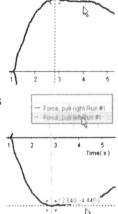

Examine the Table

- The Table shows the data for both sensors and also the sum of the data. The statistics row shows the greatest force for each sensor, and the minimum and maximum sums.

▲Force, pull right Run #1		■Force, pull left Run #1		●delta force Run #1	
Time (s)	Force pull right (N)	Time (s)	Force pull left (N)	Time (s)	Delta Force (N)
5.230	3.593	5.230	-3.594	5.230	0.001

8.   Observe how the force values compare row-by-row. Try to find a row where the force values are equal but opposite.

9.   Record the Minimum and Maximum values for the forces and for "delta force" column.

	Force, pull right (N)	Force, pull left (N)	Delta force (N)
Minimum			
Maximum			

Observations

How did the shape of the graph for the first force sensor compare to the shape of the graph for the second force sensor during the Tug-of-War?

Comparisons

1.   If two forces are equal and opposite, what should the sum of the forces be?

2.   From your data, how close are the forces to being equal but opposite?

Synthesize

Error Analysis

What were the sources of error in this experiment?

Conclusions

During a tug-of-war, how does the force on one force sensor compare to the force on the other force sensor?

Do your results support your hypothesis?

Applications

What are examples of physical events where Newton's Third Law applies?

Extension Problem

In Cutnell & Johnson, Physics, 6th ed., Volume One, Chapter 4, problem 16, page 117.

At a time when mining asteroids has become feasible, astronauts have connected a line between their 3500-kg space tug and a 6200-kg asteroid. Using their ship's engine, they pull on the asteroid with a force of 490 N. Initially the tug and the asteroid are at rest, 450 m apart. How much time does it take for the ship and the asteroid to meet?

Activity 6: Work and Energy
(Force Sensor, Photogate)

Preface

- *If* you are using the PASCO electronic Workbook specifically designed for this activity, then do the following:
1. Connect the *USB Links* to the computer's USB port.
2. Connect the *Force Sensor* to one of the USB Links and the *Photogate Port* to the other. This will automatically launch the PASPORTAL window.
3. Choose the electronic Workbook entitled: **06 Work and Energy WB.ds** and follow the directions in the Workbook.

Introduction

The purpose of this exploration is to compare the work done on an object to the change in kinetic energy of the object.

Use the Force Sensor to measure the force applied to the cart. Use the Photogate/Pulley System to measure the motion of the cart as it is pulled by the weight of the hanging mass.

Most people expect that if you do work, you get something as a result. In physics, when a net force performs work on an object, there is always a result from the effort. The result is a change in the energy of the object. The relationship that relates work to the change in kinetic energy is known as the work-energy theorem. This theorem is obtained by bringing together three basic concepts: Newton's Second Law, work, and the equations of kinematics.

Learning Outcomes

You will be able to:

- Qualitatively understand the Work-Energy Theorem.

- Make measurements to determine the amount of work done on an object.

- Use sensors to determine the change in kinetic energy of an object.

- Compare the change in kinetic energy of an object to the work done on the object (the area under a graph of force versus distance).

Hypothesis

How does the amount of work done on an object to make it move in the horizontal direction compare to the amount of kinetic energy gained by the object?

Background

For an object with mass m that experiences a net force F_{net} over a distance d that is parallel to the net force, Equation 6.1 shows the work done.

$$W = F_{net}d$$

Equation 6.1

If the work changes the object's vertical position, the object's gravitational potential energy changes.

$$W = \Delta KE = KE_f - KE_i = \frac{1}{2}mv_f^2 - \frac{1}{2}mv_i^2$$

Equation 6.2

However, if the work changes only the object's speed, the object's kinetic energy changes as shown in Equation 6.2 where W is the work, vf is the final speed of the object and vi is the initial speed of the object.

For more information see Cutnell & Johnson, Physics, 6th ed., Volume One, Chapter 6, Section 6.5.

Materials

Equipment Needed	Qty	Equipment Needed	Qty
PASPORT Force Sensor (PS-2104)	1	Mass and Hanger Set (ME-9348)	1
PASPORT Photogate Port (PS-2123)	1	Photogate/Pulley System (ME-6838)	1
USB Link (PS-2100)	2	Compact Cart Mass (ME-6755)	1
1.2 m PAScar Dynamics System (ME-6955)	1	Universal Table Clamp (ME-9376B)	1
Balance (SE-8707)	1	String (SE-8050)	1.2 m

Setup

Computer Setup

1. Plug the *USB Links* into the computer's USB port.

2. Plug the *PASPORT Force Sensor* into one of the USB Links and the *PASPORT Photogate Port* into the other. This will automatically launch the PASPORTAL window.

3. Choose the DataStudio configuration file entitled **06 Work and Energy CF.ds** and proceed with the following instructions.

SAFETY REMINDER	THINK SAFETY
• Follow the directions for using the equipment.	ACT SAFELY BE SAFE!

Equipment Setup

1. Use the thumbscrew that comes with the Force Sensor to mount the sensor onto the accessory tray of the cart. Add a 250-g mass to the cart.

2. Measure the total mass of the cart, Force Sensor, and 250-g mass. Convert the mass to kilograms and record the value.

Mass of cart, Force Sensor, and 250-g mass:

3. Place the track on a horizontal surface. Level the track by placing a cart on the track. If the cart rolls one way or the other, use the leveling screw at one end of the track to raise or lower that end until the track is level and the cart does not roll one way or the other.

4. Note: It is very important that the track is level to get the best results.

5. Use the Pulley Mounting Rod to attach the Pulley to the tab on the Photogate.

6. Attach the Universal Table Clamp on the end of the track. Mount the Photogate/Pulley System's rod in the clamp so that the top edge of the pulley is approximately the same height as the hook on the Force Sensor that is mounted on the cart.

7. Place the end stop with magnets about 1 cm in front of the smart pulley. Arrange the cart so that the magnets on the cart and the magnets on the end stop face one another.

8. Use a piece of string that is about 10 centimeters longer than the distance from the top of the pulley to the floor. Connect one end of a string to the Force Sensor's hook. Place the string in the Pulley's groove.

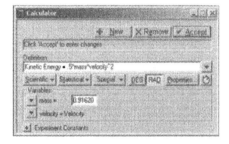

9. Attach the hanger with the known mass to the end of the string in such a way that the bottom of the hanger is just above the floor when the cart is against the end-stop.

10. Put a known mass (e.g., 50-g) on the mass hanger. Weigh the hanger and the mass. Convert the mass to kilograms and record the value.

Hanger and mass:

11. Add the masses together to find the total mass of the system. Record the total mass.

*   The calculation for Kinetic Energy uses the total mass of the system.

Total mass:

12. Click the **Calculate** button to open the Calculator window. Enter the value for the total mass and then click **Accept**. (In the example, the mass is 0.91620 kg). Close the window after you enter the mass.

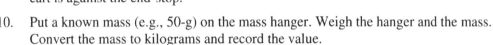

13. Press the Zero button on the Force Sensor to zero the sensor.

Record Data

*   Reminder: Stop the cart before the end stop. Push the Zero button on the Force Sensor.

1. Pull the cart away from the Photogate so the mass on the end of the string is just below the pulley.

2. Hold the Force Sensor cable so the cart can move freely. Click **Start**. Release the cart so that it can move toward the Photogate.

*   Data recording stops automatically when the cart moves 65 cm (0.65 m). The Table shows the Velocity and Kinetic Energy.

3. Record the value for the Maximum Kinetic Energy from the Table.

Maximum Kinetic Energy:

*   The next section shows how to analyze the force versus distance data.

Analyze

Examine the Graph

- The work done on the object is the area under the curve for the Force vs. Position graph.

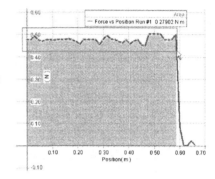

1. Click the **Scale to Fit** button to rescale the graph.

2. Highlight the flat part of the graph. (See the example.)

- The value for the Area under the curve will appear in the Legend box. (In this example, the area is 0.27902 N m.)

3. Record the Area as the Work (N m).

Work (area):

Observations

Why is the graph of the force vs. position not uniform? Explain.

Comparisons

How does the maximum kinetic energy compare to the work done (area under the curve)?

Maximum Kinetic Energy:

Work (area):

Remember, a joule is a newton-meter.

Synthesize

Verification

Compute the percent difference between the Maximum Kinetic Energy and the Work (area under the force-position graph). Use the area under the graph as the theoretical value.

$$\% \text{ Difference} = \left| \frac{\text{measured - theoretical}}{\text{theorectical}} \right| \times 100\%$$

Percent difference:

Error Analysis

What were the sources of error in this experiment?

Conclusions

1. What is the relationship between kinetic energy and work?

2. How do you know when work is being done?

3. Do your results support your hypothesis?

Applications

The rocket boosters on the sides of the Space Shuttle Endeavor supply a constant force of 14.4 million newtons each. They are used to lift the shuttle from the ground to an altitude of roughly 45.7 kilometers. How much work do these boosters do in one launch?

Note: This is about as much energy as the average American family will use in their household over the next 40 years!

Extension Problem

In Cutnell & Johnson, Physics, 6th ed., Volume One, Chapter 6, problem 12, page 175.

A fighter jet is launched from an aircraft carrier with the aid of its own engines and a steam-powered catapult. The thrust of its engines is 2.3×10^5 N. In being launched from rest it moves a distance of 87 m and has a kinetic energy of 4.5×10^7 J at lift-off. What is the work done on the jet by the catapult?

Activity 7: Conservation of Energy
(Motion Sensor)

Preface

Introduction

One of the most important ideas you will encounter in science is the principle of conservation of energy. Energy manifests itself in a myriad of forms. Gravitational potential energy, electrical potential energy, and kinetic energy are just a few forms.

Explore the principle of conservation of mechanical energy. Confirm that gravitational potential energy can be converted into kinetic energy.

Use the Motion Sensor to measure the motion of a cart after it travels down an inclined plane.

Learning Outcomes

You will be able to:

* Understand the principle of the conservation of energy.

* Apply the principle to situations involving gravitational potential energy and kinetic energy.

* Make theoretical predictions of the final velocity of a cart after it rolls down an inclined plane using the principle of the conservation of energy.

* Compare your predictions with experimental results.

Hypothesis

How does the gravitational potential energy of a cart at the top of an inclined plane compare to the kinetic energy of the cart at the bottom of the inclined plane?

How does the total energy of the cart change as it goes down an inclined plane?

Background

Scientists have concluded that the law of conservation of energy is a description of nature. That is, as long as the system under investigation is closed so that objects do not move in and out, and as long as the system is isolated from external forces, then energy can only change form. The total amount of energy is constant. In other words, energy can be neither created nor destroyed. In a closed, isolated system, energy is conserved.

There is a separate expression for each form of energy. The expression for gravitational potential energy (GPE) is shown in Equation 7.1.

$$PE_{grav} = mgh$$

Potential energy is sometimes called "energy of position". That is, an object has usable energy based purely on its location in space and its mass. As you lift an object higher and higher away from the surface of the Earth, it gains potential energy. This can be converted into kinetic energy if it is allowed to fall. If you know the mass of the object (m), the height it can fall (h) and the value of acceleration on Earth's surface due to gravity (g), you can use Equation 7.1 to find its potential energy.

Equation 7.1

The expression for kinetic energy (KE) is shown in Equation 7.2.

$$KE = \frac{1}{2}mv^2$$

Equation 7.2

Kinetic Energy can be described as "energy of motion". An object has kinetic energy if it has mass and velocity in any direction. As you accelerate an object faster and faster from rest, it gains kinetic energy. If you know an object's mass (m) and its velocity (v), you can find its total kinetic energy by using the expression above.

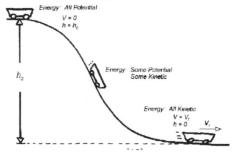

The graphical representation shows the relationship of potential and kinetic energy.

For more information see Cutnell & Johnson, Physics, 6th ed., Vol. One, Chapter 6, Section 6.5.

Materials

Equipment Needed	Qty	Equipment Needed	Qty
PASPORT Motion Sensor (PS-2103)	1	Large Rod Base (ME-8735)	1
USB Link (PS-2100)	1	IDS Adjustable Feet (ME-9470)	2 pr.
1.2 m PAScar Dynamics System (ME-6955)	2	Balance (SE-8723)	1
Meter Stick	1	Rod, 45 cm (ME-8736)	1

Setup

Computer Setup

1. Plug the *USB Link* into the computer's USB port.

2. Plug the *PASPORT Motion Sensor* into the USB Link. This will automatically launch the PASPORTAL window.

3. Choose the DataStudio configuration file entitled **07 Conserve Energy CF.ds** and proceed with the following instructions.

SAFETY REMINDER	**THINK SAFETY**
• Follow the directions for using the equipment.	**ACT SAFELY BE SAFE!**

Equipment Setup

1. Put two adjustable feet on one of the tracks; one near each end. Level this track as precisely as possible.

2. Put the remaining adjustable foot on one end of the other dynamics track. Move the pivot clamp near the other end of this track.

3. Mount the pivot clamp onto the support rod. Raise the end of the track about 15 cm and tighten the pivot clamp.

4. Place the downhill end of the inclined track against the end of the level track that does not have the fixed end stop.

5. Use the adjustable foot on the inclined track to raise or lower that end of the track until it is matched with the end of the level track.

6. Attach the Motion Sensor to the end of the level track.

7. Measure the height of the level track. Use a pencil to mark the spot on the inclined track that is exactly 2.5 cm higher than the level track.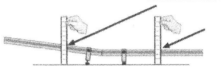

2.5 cm higher

8. Also mark spots on the inclined track that are 5 cm, 7.5 cm, and 10 cm higher than the level track.

9. Measure and record the mass of the cart (in kilograms).

10. Using the equations for the principle of conservation of energy, calculate the theoretical velocities of the cart for the following heights: 10 cm, 7.5 cm, 5 cm and 2.5 cm.

Heights (m)	Theoretical V (m/s)
0.100	
0.075	
0.050	
0.025	

Mass of cart:

Record Data

- Reminder: Stop the cart before it reaches the Motion Sensor.

1. Hold the center of mass (c.o.m.) of the cart at the spot that is 10 cm higher than the level track.

2. Click **Start**. Release the cart.

3. Catch the cart and click **Stop** just before the cart reaches the Motion Sensor.

4. For Run #2, #3 and #4, put the c.o.m. of the cart at the spot that is 7.5 cm, 5.0, and 2.5 cm higher respectively than the flat track. Repeat the process for each height.

- The next section describes how to analyze your data.

Analyze

Examine the Graph

- The graph should show the most recent data (Run #4).

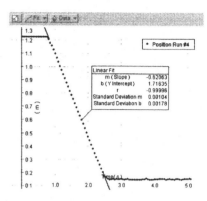

1. Click **Scale to Fit** to rescale if needed.

2. Highlight the part of the graph from about 1.0 m to the point where you stopped the cart.

3. Click the **Fit** menu and select **Linear Fit**. The slope "m" is the velocity. Record the velocity.

4. Use the **Data** menu to select Run #3.

5. Repeat the process to find the slope "m" for your other runs of data.

6. Use the procedure described previously to find the slope "m" for each run of data. Record the velocity (slope) for each run.

- Ignore the slope's negative sign. The sensor measures objects moving toward it as going in a negative direction.

Run	Height (m)	Measured V (m/s)
1	0.100	
2	0.075	
3	0.050	
4	0.025	

Observations

Did the cart's velocity decrease when it was released from the lower marks? Why?

Comparisons

1. How do the measured and theoretical values for velocity compare?

Use the mass of the cart and g = 9.8 m/s/s to calculate the theoretical gravitational potential energy (GPE) for the cart at the 10 cm (0.10 m) height. Use the measured velocity of the cart for the 10 cm height (Run #1) to calculate the kinetic energy (KE) of the cart.

Run	Height (m)	GPE (J)	KE (J)
#1	0.010		

Mass of cart:

2. How does the gravitational potential energy for Run #1 compare to the kinetic energy?

3.   Based on your results, did all of the initial gravitational potential energy convert into kinetic energy?

4.   If there is a difference between the gravitational potential energy and the kinetic energy, what do you think caused the difference?

Synthesize

Variables

1.   What was the independent variable in this activity?

2.   What did you measure?

3.   What was the result when you changed the independent variable?

Verification

Use your measured and theoretical values to compute the percent difference for each run.

$$\% \text{ Difference} = \frac{|\text{measured} - \text{theoretical}|}{\text{theorectical}} \times 100\%$$

Run	Height	Measured V (m/s)	Theoretical V (m/s)	Percent Difference
1	0.100			
2	0.075			
3	0.050			
4	0.025			

Error Analysis

What were the sources of error in this experiment?

Conclusions

Did the initial height of the cart have any effect on the data? Explain.

Do your results support your hypothesis?

Applications

In winter ski jumping, how would the skier know the final velocity at the bottom of the ramp if they are given the height of the ramp?

Extension Problem

In Cutnell & Johnson, Physics, 6th ed., Volume One, Chapter 6, problem 34, page 176.

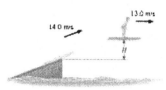

A water skier lets go of the tow rope upon leaving the end of a jump ramp at a speed of 14.0 m/s. As the drawing indicates, the skier has a speed of 13.0 m/s at the highest point of the jump. Ignoring air resistance, determine the skier's height above the top of the ramp at the highest point.

Activity 8: Impulse v Change in Momentum
(Force Sensor, Motion Sensor)

Preface

> - *If* you are using the PASCO electronic Workbook specifically designed for this activity, then do the following:
> 1. Connect the *USB Links* to the computer's USB port.
> 2. Connect the *Motion Sensor* to one USB Link and the *Force Sensor* to the other. This will automatically launch the PASPORTAL window.
> 3. Choose the electronic Workbook entitled: **08 Impulse WB.ds** and follow the directions in the Workbook.

Introduction

In this exploration you will determine the similarities between the change in momentum and the impulse (net force multiplied by time).

Use the Motion Sensor to measure the motion of a cart as it collides with a magnetic bumper.

Use a Force Sensor mounted on the track to measure the force of the collision over the same interval of time. Compare the change in momentum of the cart with the area of the measured force vs. time graph.

Learning Outcomes

You will be able to:

- Understand the relationship between impulse and change in momentum.

- Use sensors to measure the change in velocity of a cart and the force when the cart collides with a bumper.

- Calculate the change in momentum using the mass and the change in velocity.

- Calculate the impulse using the force and the time interval.

- Recognize that the area under a plot of force vs. time for a collision is equal to the impulse.

Hypothesis

How is the amount of impulse exerted on an object during a collision related to the change in momentum of the object during the collision?

Background

The impulse of a force is the product of the average force and the time interval during which the force acts. (See equation 8.1.) Impulse is a vector quantity and has the same direction as the average force. The SI unit of impulse is the newton·second (N·s).

$$\text{Impulse} = F\Delta t$$

Equation 8.1

When a net force acts on an object, the impulse of the net force is equal to the change in momentum of the object.

Impulse = Change in momentum

$$F\Delta t = mv_f - mv_i = \Delta mv$$

It is possible for the object to undergo the same change in momentum whether it is involved in an abrupt hard collision or a cushioned collision.

For more information see Cutnell & Johnson, Physics, 6th ed., Volume One, Chapter 7, Section 7.1.

Materials

Equipment Needed	Qty	Equipment Needed	Qty
PASPORT Motion Sensor (PS-2103)	1	Accessory Bracket with Bumpers (CI-6545)	1
PASPORT Force Sensor (PS-2104)	1	IDS Adjustable Feet (ME-9470)	1 pr
USB Link (PS-2100)	2	Balance (SE-8723)	1
1.2 m PAScar Dynamics System (ME-6955)	1	Heavy object, such as a book	1

Setup

Computer Setup

1. Plug the *USB Links* into the computer's USB port.

2. Plug the *PASPORT Motion Sensor* into one USB Link and the *PASPORT Force Sensor* into the other. This will automatically launch the PASPORTAL window.

3. Choose the DataStudio configuration file entitled **08 Impulse CF.ds** and proceed with the following instructions.

SAFETY REMINDER	THINK SAFETY
• Follow the directions for using the equipment.	ACT SAFELY BE SAFE!

Equipment Setup

1. Mount the Force Sensor on the Accessory Bracket. Mount the Accessory Bracket in the T-slot on the side of the Dynamics Track.

2. Attach the Adjustable Feet to each end of the track. Set one of the Adjustable Feet to its maximum height and set the other to its minimum height.

3. Place the Motion Sensor at the raised end of the track so it can measure the motion of the cart. Set the Motion Sensor Range to Cart.

• For all trials start the cart at this place.

4. Brace the Force Sensor end of the track against a heavy object so the track will not move during the collision.

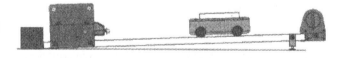

5. Place the magnetic bumper that comes with the Accessory Bracket on the front of the Force Sensor.

6.   Put the 250-g mass into the tray on the the PAScar and weigh the cart and mass together. Record this mass in kilograms.

Mass of cart:

Record Data

•   Prior to each data run, press the Zero button on the Force Sensor.

1.   Hold the cart 35 cm away from the Motion Sensor.

2.   Click **Start**. Release the cart so that it rolls forward.

•   Data recording begins when the cart collides with the bumper.

3.   Click **Stop** after the PAScar has rebounded from the collision.

•   The next section describes how to analyze your data.

Analyze

Examine the Graph: Find the Area Under the Force Curve

1.   Use the Zoom Select tool to expand a region of the force curve that shows the collision between the cart and the Force Sensor.

•   The Legend box shows the Area under the part of the curve that you selected. (In this example, Area is 0.37428 N s.). Record the Area as the Impulse (in newton seconds, N s).

Area
• Force Run #1 0.37428 N s

Impulse:

2.   Click the velocity graph. Click the **Smart Tool** button.

3.   Drag the **Smart Tool** to the point right before the curve decreases that marks the beginning of the collision.

4.   Move your cursor to the bottom right corner of the **Smart Tool**.

•   The cursor turns into a triangle (the "delta" tool).

5.   Drag the delta tool to the point at the bottom of the curve that marks the end of the collision.

•   The value for the change in the Y coordinate is the change in velocity from the beginning to end of the collision. In this example, the change is -0.68797 m/s.

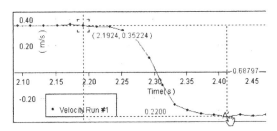

6.   Record the value for your change of velocity.

Change of velocity:

Observations

1.   Why does the velocity of the cart change from a positive value before the collision to a negative value after the collision?

2.   Why is the area under the curve for the force graph equal to the impulse?

Calculations

Calculate the change in momentum. Multiply the total mass of the cart by the change in velocity. Record the result.

Change in momentum:

Comparisons

Compare the change in momentum to the impulse under the force curve). Calculate the percent difference.

$$\% \text{ Difference} = \left| \frac{\Delta mv - \text{Impulse}}{\text{Impulse}} \right| \times 100\% \quad \text{(Area}$$

Mass (kg):	Change in velocity (m/s):	Change in momentum (kg m/s):	Impulse (N s):	% diff:

Synthesize

Error Analysis

What are possible reasons why the change in momentum is different from the measured impulse?

Conclusions

For your data, how does the change in momentum compare to the impulse?

Do your results support your hypothesis?

Applications

Explain why airbags in cars can help to prevent injuries to the occupants during a frontal collision.

Extension Problem

In Cutnell & Johnson, Physics, 6th ed., Volume One, Chapter 7, problem 3, page 200.

A 62.0-kg person, standing on a diving board, dives straight down into the water. Just before striking the water, her speed is 5.50 m/s. At a time of 1.65 s after entering the water, her speed is reduced to 1.10 m/s. What is the average net force (magnitude and direction) that acts on her when she is in the water?

Activity 9A: Conservation of Linear Momentum Part 1 –
Elastic Collision
(Motion Sensors)

Preface

- *If* you are using the PASCO electronic Workbook specifically designed for this activity, then do the following:
1. Label one Motion Sensor as "Sensor 1". Connect this *Motion Sensor* to one of the *USB Links*. Connect this USB Link to the computer's USB port. This will automatically launch the PASPORTAL window.
2. Connect the other *Motion Sensor* to a link and connect the link to a USB port.
3. Choose the electronic Workbook entitled: **09A Conserve Momentum Part 1 WB.ds** and follow the directions in the Workbook.

Introduction

The purpose of this exploration is to determine the amount of momentum before and after an elastic collision.

Use Motion Sensors to measure the motion of two carts before and after an elastic collision.

Determine the momentum for both carts before and after the collision.

Compare the total momentum of the two carts before collision to the total momentum of both carts after collision.

Learning Outcomes

You will be able to:

- Measure the velocity of carts before and after an elastic collision.

- Determine the change of momentum of the carts during the collision.

- Compare the momentum before and after the collision for each cart based on the measured masses and velocities.

Hypothesis

How does momentum before an elastic collision compare to the momentum after the collision?

Background

When objects collide, whether locomotives, shopping carts, or your foot and the sidewalk, the results can be complicated. Yet even in the most chaotic of collisions, as long as there are no net external forces acting on the colliding objects, one principle always holds and provides an excellent tool for understanding the dynamics of the collision. That principle is called the conservation of momentum. For a two-object collision, momentum conservation is easily stated mathematically by Equation 9.1.

$$m_1 v_1 + m_2 v_2 = m_1 v_1' + m_2 v_2' \qquad \textbf{Equation 9.1}$$

If net external forces are ignored, the sum of the momenta of two carts prior to a collision is the same as the sum of the momenta of the carts after the collision.

The change in momentum for each cart is its mass times its change in velocity.

$$m_1 \Delta v_1 = m_2 \Delta v_2$$

$$m_1 \left(v_1' - v_1 \right) = m_2 \left(v_2' - v_2 \right)$$

For more information see Cutnell & Johnson, Physics, 6th ed., Chapter 7, Section 7.2,7.3.

Materials

Equipment Needed	Qty	Equipment Needed	Qty
PASPORT Motion Sensor (PS-2103)	2	1.2 m PAScar Dynamics System (ME-6955)	1
USB Link (PS-2100)	1	Balance (SE-8723)	1

Setup

Computer Setup

1. Label one *PASPORT Motion Sensor* as "Sensor 1". Plug this sensor into one of the *USB Links*. Plug this link into the computer's USB port. This will automatically launch the PASPORTAL window.

2. Plug the other *PASPORT Motion Sensor* into the other link and plug the link into the USB port.

3. Choose the DataStudio configuration file entitled **09A Conserve Momentum Part 1 CF.ds** and proceed with the following instructions.

SAFETY REMINDER	THINK SAFETY
• Follow the directions for using the equipment.	ACT SAFELY BE SAFE!

Equipment Setup

•   You can use the same setup for both Part 1 and Part 2 of Conservation of Linear Momentum.

•   Make sure that magnets are installed in one end of each cart so the carts can repel each other during the collision. Label one cart "Cart 1" and label the other "Cart 2".

1. Measure the mass of each cart in kilograms and record the values.

Mass of Cart 1:

Mass of Cart 2:

2. Place the track on a horizontal surface with the fixed end stop to the left. To level the track, place a cart on the track and see if the cart rolls one way or the other. Use the leveling screw on the fixed end stop to raise or lower that end until the track is level and the cart will not roll one way or the other on its own.

*   Note: It is very important that the track is level to get the best results.

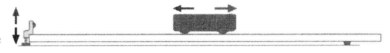

3. Set the first Motion Sensor on the left end of the track. Put the second Motion Sensor to the right end of the track. (The first Motion Sensor is the first one you connected to the computer's USB port.)

4. Adjust each sensor so it can measure the motion of a cart as it moves from the end of the track to the middle and back again. Put the Range Setting on the sensors to 'Cart'.

5. Place Cart 1 on the left side of the track and place Cart 2 on the right side. Be sure that the magnetic ends of the carts will repel each other.

*   Usually, motion away from the sensor is positive and motion toward the sensor is negative. For this exploration, a calculation is done on the data from Sensor 2 to reverse the directions. For both sensors, motion to the right is positive and motion to the left is negative.

Record Data

1. Position each cart 15 cm from the Motion Sensors.

2. Click **Start**. Gently push the carts and let them roll toward each other.

*   Continue collecting data until the carts have collided and returned to the ends of the track.

3. Click **Stop**.

*   The next section describes how to analyze your data.

Analyze

Examine the Graph: Find the Change in Velocity for Each Cart

*   Find the change of velocity from just before the collision to just after the collision.

1. Select "V Cart 1 Run #1" in the Legend box. Click the **Smart Tool** button.

2. Move the **Smart Tool** to a point on the curve where the collision begins. Move the cursor to the lower right corner of the **Smart Tool** until the "delta" appears.

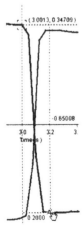

3.   Drag the "delta" Smart Tool to the point where the collision ends. Note the Delta X and the Delta Y (0.2000 and -0.65008 in the example).

4.   Select "V Cart 2 Run #1" in the Legend box. Click the **Smart Tool** again. Repeat the process to find Delta X and Delta Y (0.2000 and 0.65017 in the example).

Record your values of Delta Y (change in velocity):

Cart 1 change in velocity (m/s):

Cart 2 change in velocity (m/s):

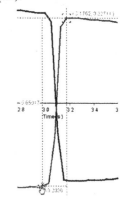

Observations

Describe the motion of the carts during the elastic collision.

Calculations

Elastic Collision: Use the mass and the change in velocity of each cart to calculate the change in momenta of Cart 1 and Cart 2.

Cart	Mass (kg)	Change in velocity (m/s)	Change in momentum (kg m/s)
1			
2			

Comparisons

How does the change in momentum of Cart 1 compare to the change in momentum of Cart 2?

Synthesize

Error Analysis

What were the sources of error in this experiment?

Conclusions

How does the total momentum before an elastic collision compare to the total momentum after the collision? Do your results support your hypothesis?

Applications

Would momentum be conserved if a car ran into a diesel train?

Extension Problem

In Cutnell & Johnson, Physics, 6th ed., Chapter 7, problem 29, page 202.

A cue ball (mass = 0.165 kg) is at rest on a frictionless pool table. The ball is hit dead center by a pool stick, which applies an impulse of +1.50 N s to the ball. The ball then slides along the table and makes an elastic head-on collision with a second ball of equal mass that is initially at rest. Find the velocity of the second ball just after it is struck.

Activity 9B: Conservation of Linear Momentum Part 2 – Inelastic Collision (Motion Sensors)

Preface

* *If* you are using the PASCO electronic Workbook specifically designed for this activity, then do the following:
1. Label one Motion Sensor as "Sensor 1". Connect this *Motion Sensor* to one of the *USB Links.* Connect this USB Link to the computer's USB port. This will automatically launch the PASPORTAL window.
2. Connect the other *Motion Sensor* to a link and connect the link to a USB port.
3. Choose the electronic Workbook entitled: **09B Conserve Momentum Part 2 WB.ds** and follow the directions in the Workbook.

Introduction

The purpose of this exploration is to determine the amount of momentum before and after an inelastic collision.

Use Motion Sensors to measure the motion of two carts before and after an inelastic collision.

Determine the momentum for both carts before and after the collision.

Compare the total momentum of the two carts before collision to the total momentum of both carts after collision.

Learning Outcomes

You will be able to:

*   Measure the velocity of carts before and after an inelastic collision.

*   Determine the change of momentum of the carts during the collision.

*   Compare the momentum before and after the collision for each cart based on the measured masses and velocities.

Hypothesis

How does momentum before an inelastic collision compare to the momentum after the collision?

If two objects with equal but opposite momenta collide head-on inelastically, what is their shared velocity after the collision?

Background

When objects collide, whether locomotives, shopping carts, or your foot and the sidewalk, the results can be complicated. Yet even in the most chaotic of collisions, as long as there are no net external forces acting on the colliding objects, one principle always holds and provides an excellent tool for understanding the dynamics of the collision. That principle is called the conservation of momentum. For a two-object collision, momentum conservation is easily stated mathematically by Equation 9.1.

$$m_1 v_1 + m_2 v_2 = m_1 v_1' + m_2 v_2' \qquad \textbf{Equation 9.1}$$

If net external forces are ignored, the sum of the momenta of two carts prior to a collision (left side of equation) is the same as the sum of the momenta of the carts after the collision (right side of equation).

For two carts in an inelastic collision, the momentum after is the product of both masses and their shared velocity.

$$m_1 v_1 + m_2 v_2 = (m_1 + m_2) v'$$

$v_1 = \text{initial V Cart 1}$

$v_2 = \text{initial V Cart 2}$

$v' = \text{shared V}$

For more information see Cutnell & Johnson, Physics, 6th ed., Volume One, Chapter 7, Section 7.2, 7.3.

Materials

Equipment Needed	Qty	Equipment Needed	Qty
PASPORT Motion Sensor (PS-2103)	2	1.2 m PAScar Dynamics System (ME-6955)	1
USB Link (PS-2100)	1	Balance (SE-8723)	1

Setup

Computer Setup

1. Label one *PASPORT Motion Sensor* as "Sensor 1". Plug this sensor into one of the *USB Links*. Plug this link into the computer's USB port. This will automatically launch the PASPORTAL window.

2. Plug the other *PASPORT Motion Sensor* into the other link and plug the link into the USB port.

3. Choose the DataStudio configuration file entitled **09B Conserve Momentum Part 2 CF.ds** and proceed with the following instructions.

SAFETY REMINDER	THINK SAFETY
• Follow the directions for using the equipment.	ACT SAFELY BE SAFE!

Equipment Setup

- You can use the same setup as for Part 1 of Conservation of Linear Momentum.

1. Label one cart "Cart 1" and label the other "Cart 2". Measure the mass of each cart in kilograms and record the values.

Mass of Cart 1:

Mass of Cart 2:

2. Place the track on a horizontal surface with the fixed end stop to the left. To level the track, place a cart on the track and see if the cart rolls one way or the other. Use the leveling screw on the fixed end stop to raise or lower that end until the track is level and the cart will not roll one way or the other on its own.

- Note: It is very important that the track is level to get the best results.

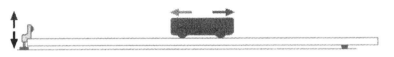

3. Set the first Motion Sensor on the left end of the track. Put the second Motion Sensor to the right end of the track. (The first Motion Sensor is the first one you connected to the computer's USB port.)

4. Adjust each sensor so it can measure the motion of a cart as it moves from the end of the track to the middle and back again. Put the Range Setting on the sensors to 'Cart'.

5. Place Cart 1 on the left side of the track and place Cart 2 on the right side. Be sure that the hook-and-pile (Velcro®) ends of the carts are facing each other.

- Usually, motion away from the sensor is positive and motion toward the sensor is negative. For this exploration, a calculation is done on the data from Sensor 2 to reverse the directions. For both sensors, motion to the right is positive and motion to the left is negative.

Record Data

1. Position each cart 15 cm from the Motion Sensors.

2. Click **Start**. Gently push the carts and let them roll toward each other.

- Continue collecting data until the carts have collided and stuck together near the middle of the track.

3. Click **Stop**.

- The next section describes how to analyze your data.

Analyze

Examine the Graph: Find the Change in Velocity for Each Cart

- Find the velocity of each cart just before impact and the shared velocity after they collide.

1. Select "V Cart 1 Run #1" in the Legend box. Click the **Smart Tool** button.

2. Use the **Smart Tool** to find the velocity of Cart 1 just before and just after the collision. Afterwards, turn off the Smart Tool.

3. Click "V Cart 2 Run #1" in the Legend box.

4. Click the **Smart Tool** and use it to find the initial velocity of Cart 2.

5. Record the velocities.

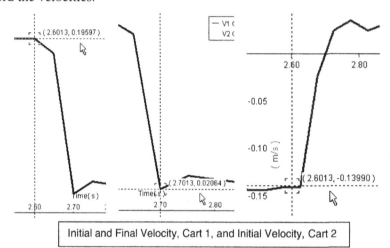

Initial and Final Velocity, Cart 1, and Initial Velocity, Cart 2

Initial V for Cart 1:

Final V for Cart 1:

Initial V for Cart 2:

Observations

What happens to the carts during the inelastic collision?

Calculations

Inelastic Collision: Use the mass and the change in velocity of each cart to calculate the momentum of each cart before collision. Find the total momentum before the collision.

Cart	Mass (kg)	Initial Velocity (m/s)	Momentum Before (kg m/s)	Total Momentum Before
1				
2				

The momentum after collision is the product of the combined mass and the velocity after collision. Use the total mass and the final velocity to determine the momentum of both carts after collision.

	Mass (kg)	Final Velocity (m/s)	Momentum After
Cart 1 + 2			

Comparisons

How does the momentum before the collision compare to the momentum after the collision?

Synthesize

Error Analysis

What were the sources of error in this experiment?

Conclusions

How does momentum before an inelastic collision compare to the momentum after the collision?

Do your results support your hypothesis?

Applications

If two objects with equal but opposite momenta collide head-on inelastically, what is their combined velocity after the collision?

Extension Problem

In Cutnell & Johnson, Physics, 6th ed., Volume One, Chapter 7, problem 25, page 202.

In a football game, a receiver is standing still, having just caught a pass. Before he can move, a tackler, running at a velocity of +4.5 m/s, grabs him. The tackler holds onto the receiver, and the two move off together with a velocity of +2.6 m/s. The mass of the tackler is 115 kg. Assuming that momentum is conserved, find the mass of the receiver.

Activity 10: Rotational Motion
(Rotary Motion Sensor)

Preface

- *If* you are using the PASCO electronic Workbook specifically designed for this activity, then do the following:
1. Connect the *USB Link* to the computer's USB port.
2. Connect the *Rotary Motion Sensor* to the USB Link. This will automatically launch the PASPORTAL window.
3. Choose the electronic Workbook entitled: **10 Rotational Motion WB.ds** and follow the directions in the Workbook.

Introduction

In this exploration you will measure the angular position and velocity of a rotating body. For each kinematic quantity (i.e. position, velocity, etc.) there is an analogous quantity in rotational kinematics. The rotational version of position (x) is the "angular position" which is given by the Greek letter theta. The rotational version of velocity (v) is "angular velocity" which is given by the Greek letter omega. All translational (linear) quantities have rotational counterparts.

Use a Rotary Motion Sensor to measure the rotation of a disk as the disk undergoes a constant angular acceleration.

Plot the angular position (angular displacement) and angular velocity and analyze them.

Compare the plots of angular position and angular velocity for the accelerating disk to plots of position and velocity for an accelerating fan cart (see PVA WB.ds).

Learning Outcomes

You will be able to:

- Describe the similarities between rotational kinematics and translational kinematics.

- Measure the angular position and angular velocity of a rotating object that is accelerating.

- Compare the angular position and angular velocity of a rotating object that is accelerating to the position and velocity of a cart that is accelerating on a track.

Hypothesis

Imagine a disk that is rotating due to a constant positive net torque. Also imagine a cart accelerating on a track due to a constant positive net force.

How will the graph of angular position versus time for the rotating disk that is accelerating look in comparison to the graph of position versus time for the cart that is accelerating on a track?

How will a graph of angular velocity versus time for the rotating disk look in comparison to the graph of velocity versus time for the cart?

Background

Rotational motion is described by using the concepts of angular displacement, angular velocity, and angular acceleration, concepts that are analogous to displacement, velocity, and acceleration when these are used for describing linear motion.

The equations of kinematics for constant linear acceleration (see Table 10.1) can be used for solving problems involving linear motion in one and two dimensions.

$$v = v_0 + at$$

$$x = \frac{1}{2}(v_0 + v)t$$

For example, the motion of a fan cart accelerating on a flat track can be described by the equations of translational kinematics.

$$x = v_0 t + \frac{1}{2}at^2$$

$$v^2 = v_0^2 + 2ax$$

The ideas of angular displacement, angular velocity, and angular acceleration can be brought together to produce a set of equations called the equations of kinematics for constant angular acceleration.

Table 10.1

The equations of kinematics for constant angular acceleration (see Table 10.2) can be used for solving problems involving rotational motion.

$$\omega = \omega_0 + \alpha t$$

$$\theta = \frac{1}{2}(\omega_0 + \omega)t$$

For example, the motion of the blades on a fan cart as they start to rotate faster and faster can be described by the equations of rotational kinematics.

$$\theta = \omega_0 t + \frac{1}{2}\alpha t^2$$

$$\omega^2 = \omega_0^2 + 2\alpha\theta$$

For more information see Cutnell & Johnson, <u>Physics</u>, 6th ed., Volume One, Chapter 8, Section 8.3.

Table 10.2

Materials

Equipment Needed	Qty	Equipment Needed	Qty
PASPORT Rotary Motion Sensor (PS-2120)	1	Large Rod Base (ME-8735)	1
USB Link (PS-2100)	1	Mass and Hanger Set (ME-9348)	1
Rotational Accessory (CI-6691)	1	String (SE-8050)	1 m
Rod, 45 cm (ME-8736)	1		

Setup

Computer Setup

1. Plug the *USB Link* into the computer's USB port.

2. Plug the *PASPORT Rotary Motion Sensor* into the USB Link. This will automatically launch the PASPORTAL window.

3. Choose the DataStudio configuration file entitled **10 Rotational Motion CF.ds** and proceed with the following instructions.

SAFETY REMINDER	THINK SAFETY
• Follow the directions for using the equipment.	**THINK SAFETY ACT SAFELY BE SAFE!**

Equipment Setup

1. Mount the Rotary Motion Sensor on a support rod.

2. Mount the Super Pulley with Table Clamp on the end of the Rotary Motion Sensor.

3.	Attach one end of a piece of thread (about 1 m) to the hole in the edge of the medium diameter part of the three-step pulley on the Rotary Motion Sensor.

4.	Adjust the Super Pulley and thread so the thread is tangent to the medium pulley.

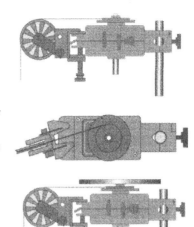

5.	Remove the thumbscrew that holds the three-step pulley onto the sensor. Place the disk of the rotational accessory on the top of the three-step pulley and replace the thumbscrew to hold the disk in place.

6.	Attach a mass hanger to the end of the thread. Turn the disk to wind the thread until the mass hanger is almost up to the Super Pulley.

7.	Hold the disk until you are ready to record data and measure the rotational motion.

Record Data

1.	When you are ready, click **Start** to begin recording data. Release the disk so it is free to rotate.

2.	Allow the disk to rotate until the string is almost completely unwound. Click **Stop** before the string unwinds all the way.

3.	Click the **Scale to fit** button to rescale the graph if needed.

•	The next section describes how to analyze your data.

Analyze

Examine the Graph

•	Compare your graph of rotational motion (angular position and velocity) to the sample graph of linear motion (position and velocity) for a cart accelerating on a track.

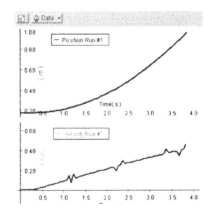

Observations

1.	What is the shape of the plot of angular position versus time?

2.	What is the shape of the plot of angular velocity versus time?

Comparisons

1.	How does the shape of the graph of angular position versus time compare to the graph of linear position versus time?

2.	How does the shape of the graph of angular velocity versus time compare to the graph of linear velocity versus time?

Synthesize

Error Analysis

What were the sources of error in this experiment?

Conclusions

Linear acceleration is determined by the ratio of the net force on an object to the mass of the object. Based on this exploration, do you think there is a similar ratio that determines the angular acceleration of a rotating object?

Do your results support your hypothesis?

Applications

What is an example of an object that first has rotational motion and then has linear motion?

Extension Problem

In Cutnell & Johnson, Physics, 6th ed., Volume One, Chapter 8, problem 19, page 226.

A flywheel has a constant angular deceleration of 2.0 rad/s/s. (a) Find the angle through which the flywheel turns as it comes to rest from an angular speed of 220 rad/s. (b) Find the time required for the flywheel to come to rest.

Activity 11: Hooke's Law
(Force Sensor, Rotary Motion Sensor)

Preface

- ***If*** you are using the PASCO electronic Workbook specifically designed for this activity, then do the following:
1. Connect the *USB Links* to the computer's USB ports.
2. Connect the *Force Sensor* to one of the USB Links and the *Rotary Motion Sensor* to the other. This will automatically launch the PASPORTAL window.
3. Choose the electronic Workbook entitled: **11 Hooke's Law WB.ds** and follow the directions in the Workbook.

Introduction

The purpose of this experiment is to demonstrate Hooke's Law both qualitatively and quantitatively.

Use a Force Sensor and a Rotary Motion Sensor to find the spring constant for a spring.

Learning Outcomes

You will be able to:

- Understand the concept of a spring constant (i.e. stiffer springs have higher spring constants).

- Use Hooke's Law to do basic calculations involving springs.

- Make measurements to calculate the spring constant for a spring.

- Compare the experimental value of the spring constant to an accepted one.

Hypothesis

What happens to the force exerted by a spring when you stretch the spring?

What makes one spring different from another?

Background

A spring that is hanging vertically from a support with no mass at the end of the spring has a length L (called its rest length). When a mass is added to the spring, its length increases by ΔL. The equilibrium position of the mass is now a distance $L + (\Delta L)$ from the spring's support. The spring exerts a restoring force, $F = -kx$, where x is the distance the spring is displaced from equilibrium and k is the force constant of the spring (also called the 'spring constant'). The negative sign indicates that the force points opposite to the direction of the displacement of the mass.

For more information see Cutnell & Johnson, Physics, 6th ed., Volume One, Chapter 10, Section 10.1.

Materials

Equipment Needed	Qty	Equipment Needed	Qty
PASPORT Force Sensor (PS-2104)	1	Harmonic Springs Set (ME-9803)	1
PASPORT Rotary Motion Sensor (PS-2120)	1	Large Rod Base (ME-8735)	1
USB Link (PS-2100)	2	Rod, 120 cm Plated, ½" dia. (ME-8741)	1
Linear Motion Accessory (CI-6688)	1	Rod, 45 cm, ½" or ¾" dia. (ME-8736)	1
Clamp, Right Angle (SE-9444)	1		

Setup

Computer Setup

1. Plug the *USB Links* into the computer's USB ports.

2. Plug the *PASPORT sensors* into the USB Links. This will automatically launch the PASPORTAL window.

3. Choose the appropriate DataStudio configuration file entitled **11 Hooke's Law CF.ds** and proceed with the following instructions.

SAFETY REMINDER	THINK SAFETY
• Follow the directions for using the equipment.	**THINK SAFETY ACT SAFELY BE SAFE!**

Equipment Setup

1. Attach the 120 cm rod to the rod stand base. Slide the Rotary Motion Sensor onto the rod. Attach the clamp, the second rod, and mount the Force Sensor vertically so its hook end is down.

2. Bend a paper clip and slide it into one of the holes in the Linear Motion Accessory.

3. Slide the Linear Motion Accessory into the Rotary Motion Sensor.

4. Attach a spring to the Force Sensor and the paper clip.

5. Separate the Rotary Motion Sensor and Force Sensor so the Linear Motion Accessory will stretch the spring when the Linear Motion Accessory is not being held.

• Note: Make sure the Force Sensor, spring, and Linear Motion Accessory are lined up vertically.

Record Data

(Hint: Read this all the way through before you begin to take data.)

1. Raise the Linear Motion Accessory so the spring is not being stretched.

2. **Zero** the Force Sensor by pressing the Zero button on the Force Sensor.

3. Click **Start**. Slowly lower the Linear Motion Accessory. The plot of the Force vs. Linear Motion appears in the graph. Click **Stop** when the paper clip hits the Rotary Motion Sensor.

- Note: It is very important to go slowly when making the measurement.

Analyze

Observations

1. Is the shape of the graph a linear function? If so, what is the relationship between the force and the stretch?

Data Analysis

To determine the spring constant, you need to find the slope of the graph.

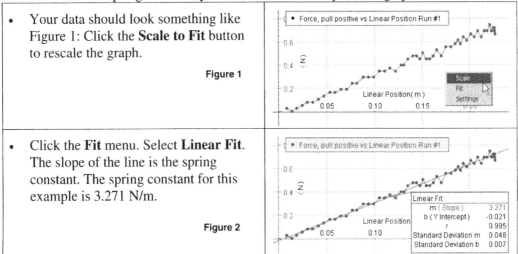

- Your data should look something like Figure 1: Click the **Scale to Fit** button to rescale the graph. **Figure 1**	
- Click the **Fit** menu. Select **Linear Fit**. The slope of the line is the spring constant. The spring constant for this example is 3.271 N/m. **Figure 2**	

Synthesize

Verification

Compare the value you obtained with your data (measured) to the value of the spring constant provided by your teacher (theoretical). Calculate the percent difference for this experiment. Record your difference below.

Percent Difference:

$$\% \text{ Difference} = \frac{|\text{measured} - \text{theoretical}|}{\text{theorectical}} \times 100\%$$

Error Analysis

What were the sources of error in this experiment?

Conclusions

What is the relationship between the force and the stretching of the spring?

Applications

Why is it important a spring scale or bathroom scale have an accurate spring constant?

Extension Problem

The following problem is from Cutnell and Johnson, Physics, 6th ed., Volume One, Chapter 10, problem 5, page 293.

A car is hauling a 92-kg trailer, to which it is connected by a spring. The spring constant is 2300 N/m. The car accelerates with an acceleration of 0.30 m/s/s. By how much does the spring stretch?

Activity 12: Simple Harmonic Motion - Mass on a Spring
(Motion Sensor)

Preface

- *If* you are using the PASCO electronic Workbook specifically designed for this activity, then do the following:
1. Connect the *USB Link* to the computer's USB port.
2. Connect the *Motion Sensor* to the USB Link. This will automatically launch the PASPORTAL window.
3. Choose the electronic Workbook entitled: **12 SHM Mass on a Spring WB.ds** and follow the directions in the Workbook.

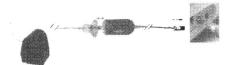

Introduction

In this exploration, you will investigate the factors that affect the period of a mass on the end of a spring.

Use the Motion Sensor to record the motion of a mass on the end of the spring.

Determine the period of oscillation and compare the value to the theoretical period of oscillation.

Learning Outcomes

You will be able to:

- Understand the basic motion of the mass on a spring and identify the factors that affect its period of oscillation.

- Determine the points in the motion at which the velocity and acceleration are minimized and maximized.

- Mathematically predict the period of oscillations for the mass on a spring.

- Measure the position of the oscillating mass.

- Determine the period of an oscillating object using position vs. time data.

- Compare the experimental period with the predicted period.

Hypothesis

What would happen to the period of oscillation for a mass on a spring if you change the mass?

How will your calculated value for the period of oscillation of the mass on the spring compare to the measured value?

Background

Note: It is a good idea to do the Hooke's Law lab before doing this one.

A spring that is hanging vertically from a support with no mass at the end of the spring has a length L (called its rest length). When a mass is added to the spring, its length increases by ΔL. The equilibrium position of the mass is now a distance $L + \Delta L$ from the spring's support. The spring exerts a restoring force, $F = -kx$, where x is the distance the spring is displaced from equilibrium and k is the force constant of the spring (also called the 'spring constant'). The negative sign indicates that the force points opposite to the direction of the displacement of the mass. The restoring force causes the mass to oscillate up and down. The period of oscillation depends on the mass and the spring constant.

For more information see Cutnell & Johnson, Physics, 6th ed., Chapter 10, Section 10.1.

Materials

Equipment Needed	Qty	Equipment Needed	Qty
PASPORT Motion Sensor (PS-2103)	1	Harmonic Spring Set (ME-9803)	1
USB Link (PS-2120)	1	Rod, 45 cm (ME-8736)	1
Mass and Hanger Set (ME-9346)	1	Large Rod Base (ME-8735)	1
Clamp, Right-Angle (SE-9444)	1	Scotch tape	5 cm
OHAUS Triple-Beam Balance (SE-8723)	1		

Setup

Computer Setup

1. Plug the *USB Link* into the computer's USB port.

2. Plug the *PASPORT Motion Sensor* into the USB Link. This will automatically launch the PASPORTAL window.

3. Choose the appropriate DataStudio configuration file entitled **12 SHM Mass on a Spring CF.ds** and proceed with the following instructions.

SAFETY REMINDER	
• Follow the directions for using the equipment.	

Equipment Setup

1. Attach one of the support rods to the Large Rod Base.

2. Weigh the spring, hanger, and a 0.050 kilogram mass. Record the value for total mass for Run #1 in kilograms.

Total mass, Run #1:

3. Using a support rod and clamp, suspend the spring so that it can move freely up-and-down. Put a mass hanger on the end of the spring.

4. Place the Motion Sensor on the floor directly beneath the mass hanger. Set the Range Select switch to the 'Person' icon.

Record Data

(Hint: Read this all the way through before you begin to take data.)

1.   Pull the mass down to stretch the spring. Release the mass. Let it oscillate a few times so the mass hanger will move up-and-down without much side-to-side motion.

2.   Click **Start**. The data will automatically stop at 3 seconds.

3.   For Run #2, add a 20-g (0.020-kg) mass to the mass hanger and record the total mass.

Total mass, Run #2:

4.   Repeat the data recording process.

•   The position curve should resemble a sine function. If it does not, check the alignment between the Motion Sensor and the bottom of the mass hanger.

Analyze

Observations

1.   What happens when the mass is pulled down a small distance from the equilibrium position and then released?

2.   What is the shape of the plot of data on your graph?

3.   Why does a spring oscillate? Explain.

Data Analysis

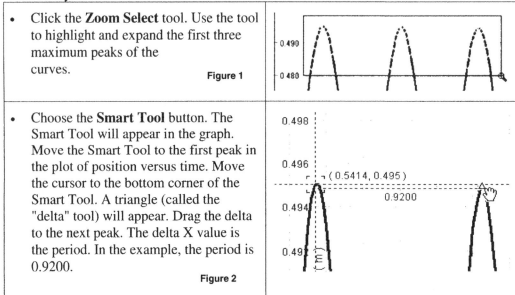

• Click the **Zoom Select** tool. Use the tool to highlight and expand the first three maximum peaks of the curves. **Figure 1**	
• Choose the **Smart Tool** button. The Smart Tool will appear in the graph. Move the Smart Tool to the first peak in the plot of position versus time. Move the cursor to the bottom corner of the Smart Tool. A triangle (called the "delta" tool) will appear. Drag the delta to the next peak. The delta X value is the period. In the example, the period is 0.9200. **Figure 2**	

Record the (measured) period for Run #1. Period 1:

Move the **Smart Tool** to the next peak. Use the delta tool to find the period between the second and third peaks. Record the period. Period 2:

Average these values and record the value. Period (avg.):

Repeat the procedure above for Run #2.

Period 1:

Period 2:

Period (avg.):

Calculations

Calculate the theoretical value of the period.

$$T = 2\pi\sqrt{\frac{m}{k}}$$

	Run #1	Run #2
Total Mass (kg):		
Spring Constant (k)*:		
Period (Theoretical):		

*Value for k either found by doing Hooke's Law lab or provided by your teacher.

Synthesize

Verification

Compare the value you obtained with your data (measured) to the value of the spring constant provided by your teacher (theoretical). Calculate the percent difference for this experiment. Record your difference.

	Run #1	Run #2
Percent difference		

$$\% \text{ Difference} = \left|\frac{\text{measured - theoretical}}{\text{theorectical}}\right| \times 100\%$$

Error Analysis

What were the sources of error in this experiment?

Conclusions

How will the period change if you increase the mass but keep the spring constant the same?

Do your results support your hypothesis?

Applications

Using your knowledge of springs, explain how a car's shock absorber works?

Extension Problem

The following problem is from Cutnell and Johnson, Physics, 6th ed., Volume One, Chapter 10, problem 21, page 293-4.

A spring stretches by 0.018 m when a 2.8-kg object is suspended from its end. How much mass should be attached to this spring so that its frequency of vibration is f = 3.0 Hz.?

Activity 13: Simple Harmonic Motion - Simple Pendulum
(Motion Sensor)

Preface

> - *If* you are using the PASCO electronic Workbook specifically designed for this activity, then do the following:
> 1. Connect the *USB Link* to the computer's USB port.
> 2. Connect the *Motion Sensor* to the USB Link. This will automatically launch the PASPORTAL window.
> 3. Choose the electronic Workbook entitled: **13 SHM Simple Pendulum WB.ds** and follow the directions in the Workbook.

Introduction

The purpose of this exploration is to determine what affects the oscillation period of a simple pendulum.

Use a Motion Sensor to measure the period of a pendulum.

Determine the relationship of the period of oscillation of a pendulum to the length of the pendulum and the mass of the pendulum.

Learning Outcomes

You will be able to:

- Understand the basic motion of the simple pendulum and what factors determine the period of oscillation.

- Mathematically predict the period of oscillations for the simple pendulum.

- Use a Motion Sensor to measure the position of the oscillating pendulum.

- Determine the period of a pendulum using position vs. time data.

- Compare the experimental period with the predicted period.

Hypothesis

What will happen to the period of oscillation of a simple pendulum when you change the length of the pendulum?

What will happen to the period of oscillation of a simple pendulum when you change the mass of the pendulum?

Background

A simple pendulum consists of a particle of mass m, attached to a frictionless pivot P by a cable of length L and negligible mass. When the particle is pulled away from its equilibrium position by an angle and released, it swings back and forth. The length L and the acceleration g due to gravity determine the frequency of a simple pendulum for small angles. The period is the reciprocal of the frequency.

$$\omega = 2\pi f = \sqrt{\frac{k}{m}}$$

For more information see Cutnell & Johnson, Physics, 6th ed., Volume One, Chapter 10, Section 10.4.

$$T = \frac{1}{f} = 2\pi \sqrt{\frac{L}{g}}$$

Materials

Equipment Needed	Qty	Equipment Needed	Qty
PASPORT Motion Sensor (PS-2103)	1	Photogate Pendulum Set (ME-8752)	1
USB Link (PS-2100)	1	Large Rod Base (ME-8735)	1
Balance (ME-8723)	1	Rod, 120 cm Plated, ½" dia. (ME-8741)	1
Clamp, Pendulum (SE-9443)	1	Braided Physics String (SE-8050)	2m
Meter Stick	1		

Setup

Computer Setup

1. Plug the *USB Link* into the computer's USB port.

2. Plug the *PASPORT Motion Sensor* into the USB Link. This will automatically launch the PASPORTAL window.

3. Choose the appropriate DataStudio configuration file entitled **13 SHM Simple Pendulum CF.ds** and proceed with the following instructions.

SAFETY REMINDER	THINK SAFETY
• Follow the directions for using the equipment.	ACT SAFELY BE SAFE!

Equipment Setup

1. Attach the 120 cm rod to the rod stand base. Attach the pendulum clamp to the rod near the top.

2. Attach a piece of string about 2 meters long to the pendulum clamp. Put the ends of the string on the inner and outer clips of the clamp so the string forms a 'V' shape as it hangs.

3. Measure and record the mass m of the first pendulum bob. Hang the first pendulum bob at the bottom of the string.

Mass: Run #1

4. Measure and record the vertical distance L from the bottom edge of the pendulum clamp to the middle of the first pendulum bob. (Note: This assumes that the center of the bob is its center of mass.)

Length: Run #1

5. Place the Motion Sensor next to the pendulum bob. Align the Motion Sensor so the brass colored disk is vertical and facing the bob and is aimed along the direction that the pendulum will swing.

6. Adjust the pendulum clamp up or down so that the pendulum bob is directly centered in front of the brass colored disk on the front of the Motion Sensor.

7. Move the sensor away from the hanging pendulum bob about 25 cm (10 in). Set the Motion Sensor to the Wide Angle (person) setting.

Record Data

(Hint: Read this all the way through before you begin to take data.)

1. Pull the first pendulum bob back about 10 cm and let it go so it can begin swinging.

2. Allow the pendulum to swing back-and-forth about 10 times to smooth its motion.

3. Click **Start**. Data recording will stop automatically after 10 seconds. (Click the **Scale to Fit** button in the graph if necessary.)

Going Further: Change the Mass

4. Remove the first pendulum bob. Measure and record the mass of the second pendulum bob and hang it on the string. (Check that the length remains the same.) Repeat the data recording process for the second pendulum bob.

Mass: Run #2

5. Measure and record the mass of the third pendulum bob and put it on the string in place of the second pendulum bob. Repeat the data recording process.

Mass: Run #3

6. Measure and record the mass of the fourth pendulum bob and put it on the string in place of the third pendulum bob. Repeat the data recording process.

Mass: Run #4

Going Further: Change the Length

7. Put the first pendulum bob back on the string. Adjust the pendulum clamp and the string to half of its original length. Measure and record the new length L. Repeat the data recording process.

Length: Run #5

8. Adjust the clamp and string again to half of the new length. Record and measure the second new length and repeat the data recording process.

Length: Run #6

Analyze

Observations

1.   How would you describe the motion of a pendulum?

2.   What is the shape of the graph of position versus time for the pendulum?

3.   What happens to the amplitude of the graph over time?

Variables

1.   What were the variables in this activity?

2.   Which of those did you control?

3.   Which did you manipulate and how did it (they) respond?

Data Analysis

To determine the period of oscillation of the pendulum bob, measure the time from one crest to the next crest (or from one trough to the next trough) for each data run.

• Use the **Data** menu in the Graph display to select Run #1 (first pendulum bob). [Hint: Select 'No Data' and then select 'Run #1.] Rescale the graph if necessary by selecting the **Scale to Fit** button. **Figure 1**	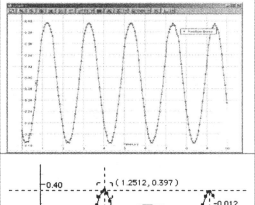
• Choose the **Smart Tool**. The Smart Tool will appear on the graph. Use the Smart Tool as described below to find the 'delta x' value for the time between crests. **Figure 2**	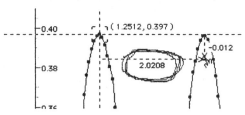

Use the Smart Tool's delta function to find the time between crests (or troughs). As shown in the example, move the Smart Tool to the first crest. Then click and drag a corner of the Smart Tool to the next crest. The 'delta x' and 'delta y' values appear on the dotted lines between the original and new positions of the Smart Tool. The period for Run #1 is circled above.

Repeat the process for the other runs of data and record the values below.

	Run #1	Run #2	Run #3	Run #4	Run #5	Run #6
Period (s)						

Synthesize

Calculations

Use the length of the pendulum to calculate the theoretical period T based on the equation and record the calculations. Also record the values for the matching measured periods.

$$T = \frac{1}{f} = 2\pi\sqrt{\frac{L}{g}}$$

Run:	#1	#5	#6
Length (m)			
Period (theo.)			
Period (meas.)			

Verification

Compare the value you obtained with your data (measured) to the value you calculated (theoretical). Calculate the percent difference for each run and record your difference in the table below.

Run:	#1	#2	#3
Percent difference			

$$\% \text{ Difference} = \left|\frac{\text{measured - theoretical}}{\text{theorectical}}\right| \times 100\%$$

1. In general, how do your measured values for the period compare to your calculated values for the period?

2. What happened to your measured values for the period when you changed the mass of the pendulum bob?

Error Analysis

What were the sources of error in this experiment?

Conclusions

What happens to the period of a pendulum if its length changes? What happens to the period of the pendulum if its mass changes?

Do the results support your hypothesis?

Applications

Give some examples where a pendulum is used for timing.

Extension Problem

The following problem is from Cutnell and Johnson, Physics, 6th ed., Volume One, Chapter 10, problem 41, page 295.

Astronauts on a distant planet set up a simple pendulum of length 1.2 m. The pendulum executes simple harmonic motion and makes 100 complete vibrations in 280 s. What is the acceleration due to gravity?

Activity 14: Buoyant Force
(Force Sensor)

Preface

- *If* you are using the PASCO electronic Workbook specifically designed for this activity, then do the following:
1. Connect the *USB Link* to the computer's USB port.
2. Connect the *Force Sensor* to the USB Link. This will automatically launch the PASPORTAL window.
3. Choose the electronic Workbook entitled: **14 Buoyant Force WB.ds** and follow the directions in the Workbook.

Introduction

Anyone who has tried to push a beach ball under the water has felt how the water pushes back with a strong upward force. This upward force is called the buoyant force and all fluids apply such a force to objects that are immersed in them. The buoyant force exists because fluid pressure is larger at greater depths.

Use the Force Sensor to measure the force on an object as it is lowered into water. Plot force versus submerged depth to obtain the density of the fluid.

Learning Outcomes

You will be able to:

- Measure the buoyant force acting on a submerged object.

- Determine the slope of a force versus depth plot.

- Apply Archimedes' Principle to determine the fluid density from the slope of the graph.

- Compare a calculated value for density with an accepted value.

Hypothesis

How does buoyant force change with depth?

In which would you feel a stronger buoyant force: a swimming pool filled with oil or with syrup?

How does the buoyant force relate to the volume and mass of the object?

Background

Archimedes' Principle states that the buoyant upward force on an object entirely or partially submerged in a fluid is equal to the weight of the fluid, mg, displaced by the object. The formula for Archimedes' Principle is shown in Equation 14.1 where ρ is the density of the fluid, V is the submerged volume of the object, and g is the acceleration due to gravity. The submerged volume is equal to the cross-sectional area, A, multiplied by the submerged height, h. So the buoyant force can be written as shown in Equation 14.2. If the object is lowered into the fluid while the buoyant force is measured, the slope of the graph of F versus h is proportional to the density of the fluid.

$$F = mg = \rho V g$$

Equation 14.1

$$F = \rho(Ah)g$$

Equation 14.2

For more information see Cutnell & Johnson, Physics, 6th ed., Chapter 11, Section 11.6.

Materials

Equipment Needed	Qty	Equipment Needed	Qty
PASPORT Force Sensor (PS-2104)	1	Density Set (ME-8569)	1
USB Link (PS-2100)	1	Large Rob Base (ME-8735)	1
Laboratory Jack, Medium (SE-9373)	1	Rod, 45 cm Plated, ½" dia. (ME-8736)	1
Clamp, Right Angle (SE-9444)	1	Braided Physics String (SE-8050)	1m
Meter Stick	1	Stainless Steel Calipers (SF-8711)	1
Beaker, 1000 ml	1	Water	500 ml

Setup

Computer Setup

1. Plug the *USB Link* into the computer's USB port.

2. Plug the *PASPORT Force Sensor* into the USB Link. This will automatically launch the PASPORTAL window.

3. Choose the appropriate DataStudio configuration file entitled **14 Buoyant Force CF.ds** and proceed with the following instructions.

SAFETY REMINDER	THINK SAFETY
• Follow the directions for using the equipment.	ACT SAFELY BE SAFE!

Equipment Setup

1. Mount the Force Sensor on a horizontal rod with the hook end down.

2. Using the calipers, measure the diameter of the metal cylinder. From the diameter, calculate the radius and the cross-section area. Record the cross-section.

$$A = \pi R^2$$

A (cross-section):

Note: It is important your value for "A" be in square meters to simplify your calculations.

3. Hang the metal cylinder from the Force Sensor hook with a string.

4. Put about 800 ml of water into the beaker and place the beaker on the lab jack below the hanging cylinder. The bottom of the cylinder should be touching the water but not submerged.

5. Position the metric ruler next to the edge of the lab jack.

6. Zero the Force Sensor before beginning to take data.

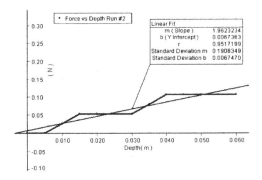

Note the initial height of the top of the lab jack.

Record Data

(Hint: Read this all the way through before you begin to take data.)

1. Click **Start**. (The Start button changes to Keep.) The Force will appear in the first cell in the Table display. Click the **Keep** button to record the first value (the force at zero depth).

2. Immerse the cylinder 5 millimeters (5mm or 0.005 m) by raising the beaker of water 5 mm with the lab jack. Use the metric ruler to measure the distance that you raised the lab jack. Click **Keep** to record the force at 5 mm.

3. Increase the depth of submersion by increments of 0.005 m. After each increase in the submersion, wait for the force reading to stabilize, then click **Keep**.

4. Click the **Stop** button when you have recorded the final value of 0.060 m. Your data will be graphed.

Analyze

Observations

1. Why was the Force Sensor zeroed after the cylinder was attached to the hook?

Variables

1. What was the independent variable in this activity (the one that you changed)?

2. What was the dependent variable (the one that responded to the change)?

3. How did the dependent variable respond?

Determine the slope of the graph.

• Click the **Scale to Fit** button to rescale the graph if needed. Click the **Fit** button. Select **Linear Fit** from the menu.

Record the Slope (m).

Slope:

Synthesize

Calculations

Calculate the density of water by setting the slope equal to (ρ)Ag and solving for ρ.

Slope (Force/Depth)	
A (cross-section)	
Acceleration of gravity (g)	
Density of water (ρ)	

$$\frac{Force}{Depth} = \rho A g$$

Verification

Compare the value you obtained with your data (measured) to the value you calculated (theoretical). Calculate the percent difference for each run and record your difference in the table below.

$$\% \text{ Difference} = \left|\frac{\text{measured - theoretical}}{\text{theorectical}}\right| \times 100\%$$

The *measured* value you found for the density of water is:

Density of water (ρ):

The *theoretical* value for the density of water is:

Density of water (ρ): 1000 kg/m^3

Compute the percent difference.

Percent difference:

How does your experimental value compare to the accepted value for the density of water? Why?

Error Analysis
What were the sources of error in this experiment?

Conclusions
How does the buoyant force in water relate to the depth?

Do the results support your hypothesis?

Applications

Why are ships weighing as much as 45,000 tons able to float? (Examples: aircraft carrier, cruise ships).

Extension Problem

The following problem is from Cutnell and Johnson, Physics, 6th ed., Volume One, Chapter 11, problem 44, page 332.

What is the smallest number of whole logs (ρ = 725 kg/m^3, radius = 0.0800 m, length = 3.00 m) that can be used to build a raft that will carry four people, each of which has a mass of 80.0 kg?

Activity 15: Temperature and Heat
(Temperature Sensor, Voltage-Current Sensor)

Preface

> • *If* you are using the PASCO electronic Workbook specifically designed for this activity, then do the following:
> 1. Connect the *USB Links* to the computer's USB ports.
> 2. Connect the *Temperature Sensor* to one of the USB Links and the *Voltage-Current Sensor* to the other. This will automatically launch the PASPORTAL window.
> 3. Choose the electronic Workbook entitled: **15 Temp and Heat WB.ds** and follow the directions in the Workbook.

Introduction

The purpose of this exploration is to study the relationship between heat, thermal energy, and temperature.

Use the Temperature Sensor to measure the temperature of 100 ml of water as a heating resistor heats it for a set amount of time. Then use the sensor to measure the temperature of 200 ml of water as the same resistor heats it for the same amount of time. Both measurements start at the same temperature.

Compare the final temperature of the 100 ml sample of water to the final temperature of the 200 ml sample of water.

Learning Outcomes

You will be able to:

- Measure the temperature of two different amounts of water.

- Determine the minimum and maximum temperatures of the two different quantities of water.

- Compare the change in temperature of one quantity of water to the change in temperature of another quantity of water when both quantities are heated by equal amounts of thermal energy.

- Compare the thermal energy gained by the smaller quantity of water to the thermal energy gained by the larger quantity of water.

Hypothesis

How will the temperature change of 100 ml of water compare to the temperature change of 200 ml of water if both quantities of water are heated equally?

In which would you feel a stronger buoyant force: a swimming pool filled with oil or with syrup?

Background

Heat is energy in transit between two or more objects. When the energy is inside an object, it is sometimes called internal energy or thermal energy. The thermal energy in an object is the total kinetic energy of all the particles that make up the object. Heat is energy in transit between two or more objects. When the energy is inside an object, it is sometimes called internal energy or thermal energy. The thermal energy in an object is the total kinetic energy of all the particles that make up the object. Greater amounts of heat are needed to raise the temperature of solids or liquids to higher values. A greater amount of heat is also required to raise the temperature of a greater mass of material. Similar comments apply when the temperature is lowered except that heat must be removed. The equation to find how much heat is supplied or removed is below.

$$Q = mc\Delta T$$

Where m is the mass of the substance, ΔT is the change in temperature, and c is the specific heat capacity of the substance. The amount of thermal energy in an object is related to temperature, but temperature by itself can't tell you how much thermal energy is in an object. For example, a bed of glowing coals in a fireplace might have a temperature of 600 °C while a single spark from the fire might have a temperature of 2000° C. The single "hot" spark gives off very little heat while the bed of relatively "cool" coals gives off a large amount of heat. The difference between the bed of coals and the single spark has to do with both the temperature and the quantity of matter. Identical thermometers in two pots of water on a hot stove will show different temperatures even if the pots have been on the stove for the same time if the amount of water in one pot is different than the amount in the other.

For more information see Cutnell & Johnson, Physics, 6th ed., Vol.1, Chapter 12, Section 12.6.

Materials

Equipment Needed	Qty	Equipment Needed	Qty
PASPORT Temperature Sensor (PS-2125)	1	Heating Resistor, 10 ohms, 1W	1
PASPORT Voltage-Current Sensor (PS-2115)	1	Graduated Cylinder, 100 ml	1
USB Link (PS-2100)	2	Styrofoam cup w/ lid	1
Power Supply, 20 V DC, 5 A (SE-9720)	1	Patch Cord (SE-9750)	1
Water	300 ml		

Setup

Computer Setup

1. Plug the *USB Links* into the computer's USB ports.

2. Plug the *PASPORT Sensors* into the USB Links. This will automatically launch the PASPORTAL window.

3. Choose the appropriate DataStudio configuration file entitled **15 Temp and Heat CF.ds** and proceed with the following instructions.

SAFETY REMINDER	THINK SAFETY
• Follow the directions for using the equipment.	ACT SAFELY
• The heating resistor gets very hot. Do not touch it while it is on.	BE SAFE!

Equipment Setup

<u>Note</u>: Be sure that the heating resistor is in the water before you turn on the power supply. Otherwise it will burn out.

1. If you have a lid that will fit over the top of the cup, make one hole in the lid for the Temperature Sensor, and a second hole in the lid for the heating resistor.

2. Put 100 ml of water in the foam cup.

3. Connect one of the banana plugs of the heating resistor into the positive outlet of the power supply.

4. Connect the other banana plug from the resistor into the positive current jack of the Voltage-Current Sensor.

5. Connect a patch cord between the negative power supply to the negative current jack on the Voltage-Current Sensor.

6. Connect the "negative" (black) voltage lead from the Voltage-Current Sensor to the negative outlet of the Power Supply.

7. Connect the "positive" (red) voltage lead from the Voltage-Current Sensor to the positive outlet of the Power Supply.

8. Insert the Temperature Sensor through one of the holes you made in the lid. Insert the heating resistor through the other hole.

<u>Note</u>: Be sure the heating resistor is submerged in water when the current is flowing through it. Otherwise it can burn up!

9. Set the DC power supply to output 10 volts at 1 amp. You can check your output using the Digits display as you record the temperature data.

10. After completing part 1, remove the Temperature Sensor and heating resistor from the cup. Pour out the 100 ml of warmed water.

11. For part 2, pour in 200 ml of water in the Styrofoam cup and perform the same exploration as described above.

Record Data

(Hint: Read this all the way through before you begin to take data.)

Part 1: Record Temperature - 100 ml

1. Click **Start**. Data recording begins when the temperature of the water reaches 20 degrees Celsius.

Note: While the power supply is ON, gently swirl the water in the cup so the water will be heated evenly.

2. Data recording stops automatically at 10 minutes. When data recording stops, turn off the DC power supply.

Part 2: Record Temperature - 200 ml

3. Set up the cup with 200 ml of water. Turn on the power supply. Click **Start**. Data recording begins when the temperature of the water reaches 20 degrees Celsius.

Note: While the power supply is ON, gently swirl the water in the cup so the water will be heated evenly.

4.   Data recording stops automatically at 10 minutes. When data recording stops, turn off the DC power supply.

Analyze

Observations

1.   How did the change in temperature of 100 ml of water compare to the change in temperature of 200 ml of water?

2.   Why is the final temperature for the 200 ml of water different than the final temperature for the 100 ml of water?

Variables

1.   What was the independent variable in this activity (what did you change)?

2.   What changed as a result of your manipulation of the independent variable? How did it change?

Data Analysis

Determine the slope of the graph.

• Click the **Scale to Fit** button.	

The Legend box shows the minimum (Min.) and maximum (Max.) temperatures for each run.

Record your data below.

	100 ml	200 ml
Temp. (max.)		
Temp. (min.)		
Change in Temp.		

Synthesize

Calculations

Calculate Q, the thermal energy gained by the water, using the mass of the water, the specific heat of water, c = 4186 J/(kg*°C), and the change in temperature. (Assume 1 ml = 1 g)

$$Q = mc\Delta T$$

	100 ml	200 ml
Change in Temp:		
Thermal Energy, Q:		

Verification

According to your calculations, did the 100 ml of water receive the same, more, or less thermal energy than the 200 ml of water? Why did this happen?

Error Analysis

What were the sources of error in this experiment?

Conclusions

Predict the outcome if you did the experiment using 300 ml of water.

Do the results support your hypothesis?

Applications

Which would heat up more rapidly given the same amount of thermal energy, 1,000 L or 10,000 L of water?

Extension Problem

The following problem is from Cutnell and Johnson, Physics, 6th ed., Volume One, Chapter 12, problem 44, page 368.

When you take a bath, how many kilograms of hot water (49.0 C) must you mix with cold water (13.0 C) so that the temperature of the bath is 36.0 C? The total mass of the water (hot plus cold) is 191 kg. Ignore any heat flow between the water and its external surroundings.

Activity 16: Specific Heat
(Temperature Sensor)

Preface

> - *If* you are using the PASCO electronic Workbook specifically designed for this activity, then do the following:
> 1. Connect the *USB Link* to the computer's USB port.
> 2. Connect the *Temperature Sensor* to the USB Link. This will automatically launch the PASPORTAL window.
> 3. Choose the electronic Workbook entitled: **16 Specific Heat WB.ds** and follow the directions in the Workbook.

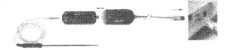

Introduction

The amount of thermal energy that an object must absorb to change its temperature by one degree is called its heat capacity. The amount of thermal energy that a single gram of a specific material must absorb in order to change its temperature by one degree is the material's specific heat capacity, or specific heat. The specific heat of water is a standard to which specific heats of other substances are compared.

In this exploration, you will measure the change in temperature of a known quantity of water at room temperature when a metal object of known mass and known initial temperature of 0 ° Celsius is put into the water.

Using your measurements of the initial and final temperatures of the water, determine the specific heat of the metal object and identity of the metal.

Learning Outcomes

You will be able to:

- Understand how specific heat can be used to find the identity of an unknown metal.

- Understand how temperature plays a key role in the idea of specific heat.

- Calculate the thermal energy absorbed by the metal object (and given up by the water.)

- Compare your experimental value for specific heat to the accepted value.

Hypothesis

How close will your experimental value for the specific heat of the unknown object be to the accepted value?

Background

When heat flows into an object, its thermal energy increases, and so does its temperature. The amount of increase depends on the size of the object. It also depends on the material from which the object is made. The specific heat of a material is the amount of energy that must be added to the material to raise the

$$Q_m = m_m C_m \Delta T_m$$

Equation 16. 1

$$Q_w = m_w C_w \Delta T_w$$

Equation 16. 2

temperature of a unit mass one-temperature unit. You will use a calorimeter to find the specific heat of the material. Since the calorimeter is relatively well insulated, the air outside will have little to do with the experiment. Inside the calorimeter, thermal energy is conserved. What this means is whatever heat is gained by the mass from the water is exactly equal to the heat lost by the water to the mass. The expressions for heat gained or lost are shown. The first equation is for an unknown metal and the second equation is for water.

Since thermal energy is conserved:

$$Q_m = -Q_w$$
$$m_m C_m \Delta T_m = -(m_w C_w \Delta T_w)$$

m_m = mass of metal

C_m = specific heat of metal

ΔT_m = change of temperature of metal

m_w = mass of water

C_w = specific heat of water

ΔT_w = change of temperature of water

The only unknown quantity in the bottom equation is the specific heat of the metal. The specific heat of water is known.

Note: The negative sign in front of the right hand equations will be canceled because the temperature change of the water will be negative.

For more information see Cutnell & Johnson, Physics, 6th ed., Vol. 1, Chapter 12, Section 12.7.

Materials

Equipment Needed	Qty	Equipment Needed	Qty
PASPORT Temperature Sensor (PS-2125)	1	Beaker, 500 ml	1
USB Link (PS-2100)	1	Calorimeter (foam cup w/ lid)	1
Mass and Hanger Set (ME-9348)	1	Graduated Cylinder, 100 ml	1
Balance (SE-8723)	1	Ice	1 cup
Braided Physics String (SE-8050)	1	Water	300 ml

Setup

Computer Setup

1. Plug the *USB Link* into the computer's USB port.

2. Plug the *PASPORT Temperature Sensor* into the USB Link. This will automatically launch the PASPORTAL window.

3. Choose the appropriate DataStudio configuration file entitled **16 Specific Heat CF.ds** and proceed with the following instructions.

SAFETY REMINDER	
• Follow the directions for using the equipment.	

Equipment Setup

1. Use a 100-g mass from the Mass and Hanger Set. Measure and record the mass of your object in grams.

2. Fill a 500 ml beaker with ice and water.

3. Tie a string to the object. Place the unknown metal in the ice-water bath for 10 minutes to cool down. Add more ice as the ice begins to melt.

4. Prepare a known quantity of water at about 25 °C. Measure the weight of the foam cup (calorimeter). Use a graduated cylinder to measure 100 ml of water that is at room temperature and put this water into the foam cup (the calorimeter). Measure the weight of the cup with the water. Subtract the weight of the cup to determine the mass of the water. Record the mass below.

Mass of metal:	
Mass of water:	

5. Put the Temperature Sensor into the ice-water bath with the object. Let the sensor equalize with the ice-water bath while the metal object is cooling.

Record Data

(Hint: Read this all the way through before you begin to take data.)

Measure Initial Temperature of the Ice-Water Bath

1. After the object has cooled for ten minutes, begin to record data. Stir the ice-water bath while you record data.

2. Select **Monitor Data** from the **Experiment** menu.

Note: It is very important to stir the water so the temperature is uniform throughout the beaker.

3. When the temperature equalizes, record the value in the text box provided.

4. Click **Stop**.

Initial temperature of metal

Measure Initial Temperature of the Calorimeter Water

5.   Move the Temperature Sensor to the calorimeter water.

6.   Wait two minutes to allow the Temperature Sensor to equalize with the calorimeter water. Select **Monitor Data** from the **Experimental** menu.

Note: Stir the water so the temperature is even throughout the calorimeter.

7.   When the temperature value in the Digits display stops changing, record the value in the text box provided.

8.   Click **Stop**.

Initial temperature of the water:

Record Temperature

9.   Lift the metal object out of the ice water and dry any droplets of water still on the object.

10.   Click **Start**. Put the cold metal object into the calorimeter. Remember to stir the water! Data recording stops automatically at 2 minutes.

Note: It is very important to stir the water so the temperature is uniform throughout the calorimeter.

Analyze

Observations

1.   Describe, in words, what happened to the temperature on the graph.

Data Analysis

To determine the spring constant, you need to find the slope of the graph.

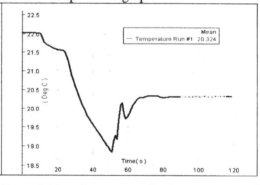

• Click the Scale to Fit button to rescale the graph if needed. Highlight the flat part of the graph.

Figure 1

The mean temperature appears in the legend. Record this value as your final temperature of metal and water.

Final Temperature:

Calculations

Fill in the table and calculate the specific heat of the unknown metal.

Mass of metal:	
Mass of water:	
C_w of water:	4.180 J/(g•°C)
Initial Temp of water:	
Initial Temp of metal:	
Final Temp of water and metal:	
Specific heat of unknown metal (C_m)	

$$m_m C_m \Delta T_m = -(m_w C_w \Delta T_w)$$

Note: Convert your value of specific heat to kilograms by multiplying your value by 1000.

Synthesize

Comparison

The table shows some specific heats of common materials.
Find the value that closely matches your data.
Enter the value and name below.

Unknown Metal	
Specific Heat (theoretical)	

Verification

Compare the value you obtained with your data (measured) to the value in the list (theoretical). Calculate the percent difference for this experiment.

$$\% \text{ Difference} = \left| \frac{\text{measured - theoretical}}{\text{theorectical}} \right| \times 100\%$$

Percent difference:

Error Analysis

What were the sources of error in this experiment?

aluminum	901 J/(kg*°C)
brass	380 J/(kg*°C)
cement	880 J/(kg*°C)
copper	386 J/(kg*°C)
glass	837 J/(kg*°C)
gold	129 J/(kg*°C)
granite	790 J/(kg*°C)
graphite	720 J/(kg*°C)
ice (-5°C)	2075 J/(kg*°C)
iron	449 J/(kg*°C)
lead	128 J/(kg*°C)
marble	860 J/(kg*°C)
silver	234 J/(kg*°C)
steam	2030 J/(kg*°C)
steel	450 J/(kg*°C)
water (15°C)	4186 J/(kg*°C)
wood	1700 J/(kg*°C)

Conclusions

What can you conclude if the specific heat you find is nowhere near a known value?

Do your results support your hypothesis?

Applications

A foundry worker throws a hot, freshly cast piece of iron into 10 liters (10 kg) of water to cool off. The initial temperature of the water is 25°C and the initial temperature of the iron is 100°C. The final temperature of the water and iron is 28°C. What is the mass of the iron? Enter your answer below.

Extension Problem

In Cutnell & Johnson, Physics, 6th ed., Volume One, Chapter 12, problem 40, page 368.

A piece of glass has a temperature of 83.0 degrees C. Liquid that has a temperature of 43.0 degrees C is poured over the glass completely covering it, and the temperature at equilibrium is 53.0. The mass of the glass and the liquid are the same. Ignoring the container that holds the glass and liquid and assuming that the heat loss to or gained from the surroundings is negligible, determine the specific heat capacity of the liquid. (Specific heat capacity of glass = 837 J/kg C.)

Activity 17: Ideal Gas Law
(Temperature Sensor, Absolute Pressure Sensor)

Preface

- *If* you are using the PASCO electronic Workbook specifically designed for this activity, then do the following:
1. Connect the *USB Links* to the computer's USB ports.
2. Connect the *Temperature Sensor* to one of the USB Links and the *Absolute Pressure Sensor* to the other. This will automatically launch the PASPORTAL window.
3. Choose the electronic Workbook entitled: **17 Ideal Gas Law WB.ds** and follow the directions in the Workbook.

Introduction

The purpose of this exploration is to study the Ideal Gas Law. You will be able to determine the relationship between pressure, volume, and temperature.

Use the Pressure Sensor to measure the pressure inside a flask and use the Temperature Sensor to measure the temperature of the water bath in which the flask is immersed.

Plot the pressure-temperature data onto a graph. Use the graph to determine the relationship of pressure and temperature and to estimate the value of Absolute Zero.

Learning Outcomes

You will be able to:

- Demonstrate that the pressure of a gas at a fixed volume is proportional to the temperature.

- Investigate the dependence of pressure on temperature.

- Use the data to determine a value for Absolute Zero and compare the experimental value to the accepted value (-273 °C).

Hypothesis

What is the relationship between the pressure of a gas and the temperature of a gas if its volume remains constant as the temperature changes?

How would you use this relationship to determine the value of Absolute Zero, the theoretical limit of low temperature?

Background

Solid, liquid and gas are the most common states of matter found on this planet. The only difference among all these states is the amount of movement of the particles that make up the substance. Temperature is a measure of the relative movement of particles in a substance because temperature is a measure of the average kinetic energy of the particles. At any specific temperature the total kinetic energy is constant. Particles with a large kinetic energy tend to collide frequently and move apart. Intermolecular forces tend to pull particles toward each other. The forces that bind some molecules together at a particular temperature are greater than the kinetic energy of the molecules. In an "Ideal Gas" there are NO intermolecular forces. (In fact, the "Ideal Gas" has no mass and occupies no volume!) While the "Ideal Gas" is fictional, real gases at room temperature and pressure behave as if their molecules were ideal. It is only at high pressures or low temperatures that intermolecular forces overcome the kinetic energy of molecules and the molecules can "grab onto" one another.

An ideal gas is an idealized model for real gases that have sufficiently low densities. The condition of low density means that the molecules of the gas are so far apart that they do not interact (except during collisions that are effectively elastic.) The ideal gas law, Equation 17.1, expresses the relationship between the absolute pressure, the Kelvin temperature, the volume, and the number of moles of the gas. In the "Ideal Gas", the volume of the gas is inversely proportional to the pressure on the gas at a constant temperature. In other words, the product of the volume and pressure for the gas is a constant when the gas is at a constant temperature. At the same time, the volume of a gas is directly proportional to the temperature. If a gas is heated, the volume of the gas increases. If it is cooled, the volume of the gas decreases.

$$PV = nRT$$

Equation 17.1

For more information see Cutnell & Johnson, Physics, 6th ed., Vol. 1, Chapter 14, Section 14.2.

Materials

Equipment Needed	Qty	Equipment Needed	Qty
PASPORT Temperature Sensor (PS-2125)	1	Aluminum Can (650-055)	1
PASPORT Absolute Pressure Sensor (PS-2107)	1	Hot Plate (SE-8767)	1
USB Link (PS-2100)	2	Rubber Stopper, one-hole	1
Beaker, 1000 ml	1	Tongs	1
Ice, crushed	1 L	Glycerin	1 ml
Water	3 L	Rubber Band	1

Setup

Computer Setup

1. Plug the *USB Links* into the computer's USB ports.

2. Plug the *PASPORT Sensors* into the USB Links. This will automatically launch the PASPORTAL window.

3. Choose the DataStudio configuration file entitled **17 Ideal Gas Law CF.ds** and proceed with the following instructions.

SAFETY REMINDER	THINK SAFETY
• Follow the directions for using the equipment. • Be careful not to touch the beaker, the boiling water, or the hot plate.	ACT SAFELY BE SAFE!

Equipment Setup

1. Put a drop of glycerin on the barb end of a quick release coupling. Put the end of the coupling into one end of a piece of plastic tubing (about 15 cm) that comes with the Pressure Sensor.

2. Put a drop of glycerin on the barb end of the connector. Push the barb end of the connector into the other end of the tubing. Fit the end of the connector into the one-hole rubber stopper.

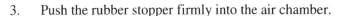
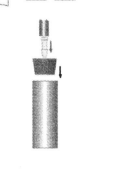

3. Push the rubber stopper firmly into the air chamber.

4. Align the quick-release coupling with the pressure port of the Pressure Sensor. Push the coupling onto the port, and then turn the coupling clockwise until it clicks (about one-eighth turn).

5. Fill the beaker with 500 ml of ice. Add 300 ml of cold water.

6. Use a rubber band to attach the Temperature Sensor to the air chamber.

7. Set the air chamber/Temperature Sensor into the ice-water bath. Use tongs to hold it in place.

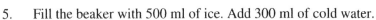

Note: Do not allow the rubber stopper to get wet.

8. Set the beaker/air chamber assembly on the hot plate.

Record Data

(Hint: Read this all the way through before you begin to take data.)

1. Click **Start**. Let the timer run for 2 minutes and then remove the ice from the beaker.

2. Do not turn off the timer while removing the ice.

3. When you have removed all of the ice, turn on the hot plate to high.

4. Click **Stop** when the water begins to boil or when the rubber stopper pops out of the air chamber. Turn off the hot plate.

Analyze

Observations

1. Is the relationship between the pressure of a gas and the temperature a linear relationship when the volume is constant?

Data Analysis

- Select the graph. Choose the **Scale to Fit** button. Highlight the top tail end of the graph Choose the **Fit** button. From the list choose **Linear**. Insert the Slope and Y-Intercept in the table below.

- Highlight the same portion of the graph as you did on the previous graph. Click the **Fit** button. From the list choose **Linear Fit**. Resize the graph until you can see where the Linear Fit line crosses the x-axis. Use the Smart Tool to find the coordinates of the point where the Linear Fit line crosses the x-axis. Record the X-intercept value (the x-coordinate) in the table below.

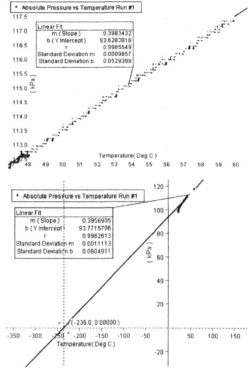

Slope:	
Y-Intercept:	
X-Intercept:	

Calculations

Calculate the X-Intercept using the equation of a line, Y-Intercept, and the slope you found previously.

Slope:	
Y-Intercept:	
X-Intercept (calc.):	

$$y = mx + b$$

The X-Intercept is the value of x when y is zero.

How does the calculated X-Intercept value compare to the X-Intercept you found on the graph? Why are they not the same?

Take the average of the X-Intercepts you found and enter the value below.

Avg. X-Intercept (measured):

Synthesize

Verification

Compare the value you obtained with your data (measured) to the accepted value (theoretical).
Calculate the percent difference for this experiment.
Record your difference below.

$$\% \text{ Difference} = \frac{|\text{measured - theoretical}|}{\text{theorectical}} \times 100\%$$

Accepted Theoretical Value: -273.14 °C

Percent Difference:

Change the value you found for Average X-Intercept to Kelvin's. $K = °C + 273.15$

K:

How does your value of absolute zero compare to the accepted value (0 K)?

Error Analysis

What were the sources of error in this experiment?

Conclusions

What happens to pressure as temperature increases?

Do your results support your hypothesis?

Applications

Is the Ideal Gas Law possible in a non-controlled environment? Why or why not?

Extension Problem

The following problem is from Cutnell and Johnson, Physics, 6th ed., Volume One, Chapter 14, problem 9, page 412.

A Goodyear blimp typically contains 5400m^3 of helium (He) at an absolute pressure of 1.1×10^5 Pa. The temperature of the helium is 280 K. What is the mass (in kg) of the helium in the blimp?

Activity 18: Superposition
(WAVEPORT Software)

Preface

Introduction

The purpose of this exploration is to study the phenomenon of superposition of sound waves. When two or more waves meet at the same place at the same time, they interfere. The combining of the waves is called superposition.

Confirm that two transverse waves of equal heights and frequency that are in phase form a shape that is the sum of the shapes of the individual waves.

Confirm that two transverse waves of equal heights and frequency that are out of phase form a shape that is the subtraction of the shapes of the individual waves.

Learning Outcomes

You will be able to:

- Describe the principle of linear superposition.

- Use WAVEPORT software to combine two sound waves so that they interfere constructively.

- Use WAVEPORT software to combine two sound waves so that they interfere destructively.

Hypothesis

What will happen when two waves of the same amplitude and wavelength meet at the same place and the same time and they are in phase?

What will happen if they meet but they are out of phase?

Background

Often, two or more sound waves are present at the same place at the same time, such as when everyone is talking at a party or when music plays from the speakers of a stereo system.

To illustrate what happens when several waves pass simultaneously through the same region, consider two transverse pulses of equal heights moving toward each other along a Slinky. In Figure 18.1 both pulses are "up." The first part of each drawing shows the two pulses beginning to overlap. The pulses merge, and the Slinky assumes a shape that is the sum of the shapes of the individual pulses. Thus, when the two "up" pulses overlap completely, as in the middle part of Figure 18.1, the Slinky has a pulse height that is twice the height of an individual pulse.

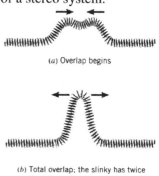

(a) Overlap begins

(b) Total overlap; the slinky has twice the height of either pulse

The two pulses move apart after overlapping, and the Slinky once again conforms to the shapes of the individual pulses.

In Figure 18.2, one pulse is "up" and the other pulse is "down." The first part of the drawing shows the two pulses beginning to overlap. The pulses merge, and the Slinky assumes a shape that is the sum of the shapes of the individual pulses. When the "up" pulse and the "down" pulse overlap exactly, as in the middle part of Figure 18.2, they momentarily cancel, and the Slinky becomes straight.

(c) The receding pulses

The two pulses move apart after overlapping, and the Slinky once again conforms to the shapes of the individual pulses.

The adding together of individual pulses to form a resultant pulse is an example of a more general concept called the principle of linear superposition.

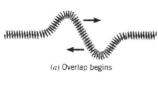

(a) Overlap begins

Suppose that the sounds from two speakers overlap in the middle of a listening area and that each speaker produces a sound wave of the same amplitude and frequency. In addition, assume the diaphragms of the speakers vibrate in phase; that is, they move outward together and inward together.

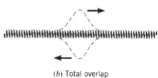

(b) Total overlap

If the distance of each speaker from the overlap point is the same, the condensations (C) of one wave always meet the condensations of the other when the waves come together.

When two waves always meet condensation-to-condensation and rarefaction-to-rarefaction (or crest-to-crest and trough-to-trough), they are said to be exactly in phase and to exhibit constructive interference.

(c) The receding pulses

Now consider what happens if one of the speakers is moved away from the overlap point by a distance equal to one-half of the wavelength of the sound. Therefore, at the overlap point, a condensation arriving from the left meets a rarefaction arriving from the right. Likewise, a rarefaction arriving from the left meets a condensation arriving from the right.

According to the principle of linear superposition, the net effect is a mutual cancellation of the two waves. The condensations from one wave offset the rarefactions from the other, leaving only a constant air pressure.

When two waves always meet condensation-to-rarefaction (or crest-to-trough), they are said to be exactly out of phase and to exhibit destructive interference.

For more information see Cutnell & Johnson, Physics, 6th ed., Vol. 1, Chapter 17, Section 17.1.

Materials

WAVEPORT Software

Setup

Computer Setup

1. Start the DataStudio program.

2. From within DataStudio, open the configuration file entitled **18 Superposition CF.ds** and proceed with the following instructions.

• *Note*: See the WAVEPORT Getting Started Guide for instructions on how to install the WAVEPORT software.

Exploration Setup

• The top half of the WAVEPORT Sound Creator window shows two Tone Select buttons, and three 'tools': the Phase Tool, Amplitude Tool, and Period Tool.

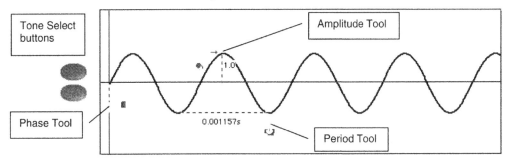

• Each tool is shaped like a hand that can be used to adjust the waveform.

• The bottom half of the WAVEPORT Sound Creator window shows the output waveform, the Speaker Tool, the Tone Control Buttons, and the Time Scale.

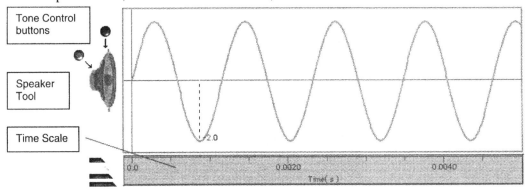

- Use the Speaker Tool to start and stop the sound output. Use the Tone Control Buttons to select which tones are added into the output waveform.

Record Data

Superposition: Two Tones

1. Click the Window menu and select the Sound Creator display labeled **Superposition Two Tones**.

2. Click the **Speaker Tool** to start the sound output.

3. Click and drag the **Phase Tool** to shift one of the waves to the right.

4. Observe what happens to the output sound when the phase of one wave is shifted relative to the other.

- Note: Click the **Speaker Tool** again to stop the sound output.

Superposition: Four Tones

1. Click the Window menu and select the Sound Creator display labeled **Superposition Four Tones**.

2. Click the **Speaker Tool** to start.

3. Click the red **Tone Select Button**. Drag the **Phase Tool** to shift the red waveform.

4. Repeat for the other tones.

5. Observe what happens to the output sound as the phase of each wave is shifted relative to the others.

- Note: Click the **Speaker Tool** again to stop the sound output.

Analyze

Observations

1. What is the shape of the output sound waveform when the two waves are in phase?

2. What is the shape of the output sound waveform when two waves are out of phase?

3. With four waveforms, what must you do to achieve destructive interference?

Synthesize

Variables

1. What were the variables in this activity?

2. Which of those did you control?

3. Which did you manipulate?

Error Analysis

What were the sources of error in this experiment?

Conclusions

What is the relationship of the phase of two sound waves and the type of interference they experience?

Do your results support your hypothesis?

Applications

How can the principle of linear superposition explain noise-canceling headphones?

Extension Problem

In Cutnell & Johnson, Physics, 6th ed., Volume One, Chapter 17, problem 9, page 507.

Speakers A and B are vibrating in phase. They are directly facing each other, are 7.80 m apart, and are each playing a 73.0 - Hz tone. The speed of sound is 343 m/s. On the line between the speakers there are three points where constructive interference occurs. What are the distances of these three points from speaker A?

Activity 19: Interference in Sound - Beats
(WAVEPORT Software)

Preface

- *If* you are using the PASCO electronic Workbook specifically designed for this activity, then do the following:
1. Open the electronic Workbook entitled: **19 Interference in Sound WB.ds** and follow the directions in the Workbook.

Introduction

The purpose of this exploration is to measure and analyze the behavior of two sounds that combine to produce beats. This activity also examines the relationship between the beat frequency and the frequencies of the two interfering sound waves.

Confirm that two sound waves of slightly different frequency will produce beats.

Confirm that the beat frequency between two sound waves of slightly different frequency is the difference in their frequencies.

Learning Outcomes

You will be able to:

- Use the WAVEPORT plug-in with DataStudio to generate two sounds with slightly different frequencies.

- Use the WAVEPORT software to determine the period of the beat frequency of the two sounds.

- Compare the beat frequency to the difference of the frequencies of the two individual sounds.

Hypothesis

What will happen when two sound waves of slightly different frequencies occur at the same time?

How will the frequency of the overall interference between the two waves compare to the difference of their individual frequencies?

Background

In situations where waves with the same frequency overlap, the principle of linear superposition leads to constructive and destructive interference.

We will see in this exploration that two overlapping waves with slightly different frequencies give rise to the phenomenon of beats. The principle of linear superposition again provides an explanation of what happens when the waves overlap.

A tuning fork has the property of producing a single-frequency sound wave when struck with a sharp blow. Figure 19.1 shows sound waves coming from two tuning forks placed side by side. The tuning forks in the drawing are identical, and each is designed to produce a 440-Hz tone. However, a small piece of putty has been attached to one fork, whose frequency is lowered to 438 Hz because of the added mass. When the forks are sounded simultaneously, the loudness of the resulting sound rises and falls periodically faint, then loud, then faint, then loud, and so on.

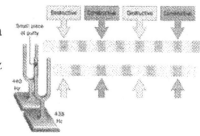

The periodic variations in loudness are called beats and result from the interference between two sound waves with slightly different frequencies.

The waves produced by the two tuning forks spread out and overlap. In accord with the principle of linear superposition, the ear detects the combined total of the two. The number of times per second that the loudness rises and falls is the beat frequency and is the difference between the two sound frequencies. For the two tuning forks, an observer hears the sound loudness rise and fall at the rate of 2 times per second (440 Hz - 438 Hz).

Figure 19.2 helps to explain why the beat frequency is the difference between the two frequencies. The drawing displays graphical representations of the pressure patterns of a 10-Hz wave and a 12-Hz wave, along with the pressure pattern that results when the two overlap.

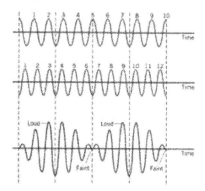

The amplitude in the red drawing is not constant, as it is in the individual waves. Instead, the amplitude changes from a minimum to a maximum, back to a minimum, and so on. They produce a loud sound when the amplitude is a maximum and a faint sound when the amplitude is a minimum.

Two loud-faint cycles, or beats, occur in the one-second interval shown in the drawing, corresponding to a beat frequency of 2 Hz.

Figure 19.2

Thus, the beat frequency is the difference between the frequencies of the individual waves, or 12 Hz - 10 Hz = 2 Hz.

For more information see Cutnell & Johnson, Physics, 6th ed., Vol. 1, Chapter 17, Section 17.4.

Materials

WAVEPORT Software

Setup

Computer Setup

1. Start the DataStudio program.

2. From within DataStudio, open the configuration file entitled **19 Interference in Sound CF.ds** and proceed with the following instructions.

•   *Note*: See the WAVEPORT Getting Started Guide for instructions on how to install the WAVEPORT software.

Exploration Setup

•   The top half of the WAVEPORT Sound Creator window shows the Frequency Control bar, two Tone Select buttons, and three 'tools': the Phase Tool, Amplitude Tool, and Period Tool.

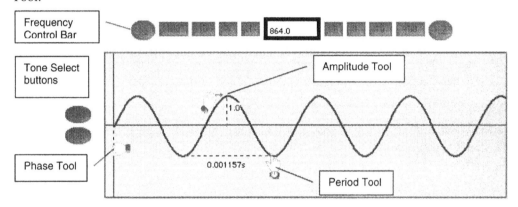

•   Each tool is shaped like a hand that can be used to adjust the waveform.

•   The bottom half of the WAVEPORT Sound Creator window shows the output waveform, the Speaker Tool, the Tone Control Buttons, and the Time Scale.

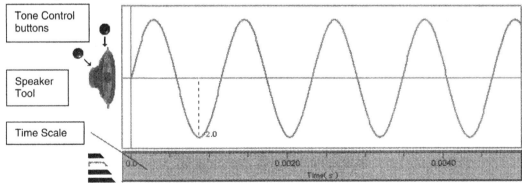

•   Use the Speaker Tool to start and stop the sound output. Use the Tone Control Buttons to select which tones are added into the output waveform.

Record Data

Interference in Sound

- The Sound Creator display is labeled **Interference in Sound**. It has two tones set at 440 Hz and 444 Hz respectively.

1. Click the **Speaker Tool** to start the sound output.

2. Use the frequency buttons to change the frequency of either the red or the green tone.

3. Observe what happens to the output sound when the frequency of one wave is changed slightly.

- Note: Click the **Speaker Tool** again to stop the sound.

4. For this part, use the frequency buttons to set the frequency of the red tone to 440.0 Hz and set the frequency of the green tone to 444.0 Hz. The frequency buttons on the right side increase the frequency and the buttons on the left side decrease the frequency.

5. After you set the frequencies, adjust the horizontal axis scale so it shows about 1 second of time. To do this, click on a number on the horizontal axis and drag it to the left. Continue to click and drag until the time at the right end of the axis is slightly more than 1.0 seconds.

6. Click the **Speaker Tool** to start the sound.

7. Click and drag the horizontal axis scale until it shows a time slightly over 1.0 seconds. Click the **Speaker Tool** to stop the sound.

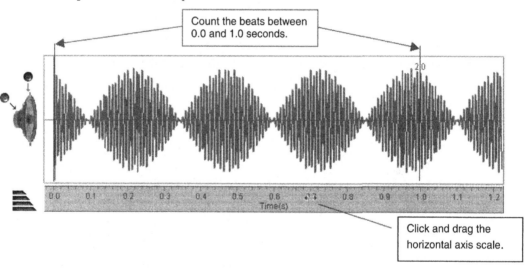

8. Count the number of beats between 0.0 and 1.0 seconds.

9. Record the number of beats.

Analyze

Calculations

What was the number of beats in one second?

How does the beat frequency compare to the difference in frequency between the two tones you used?

Observations

1. What is the shape of the output sound waveform when the two waves have a slight difference in frequency?

2. What do you hear when the two waves have a slightly different frequency?

Synthesize

Variables

1. What were the variables in this activity?

2. Which of those did you control?

3. Which did you manipulate and how did they respond?

Error Analysis

What were the sources of error in this experiment?

Conclusions

What is the relationship of the frequency of two sound waves and the type of interference they produce?

Do your results support your hypothesis?

Applications

How can the principle of beat frequency be used to tune musical instruments?

Extension Problem

In Cutnell & Johnson, Physics, 6th ed., Volume One, Chapter 17, problem 18, page 508.

A 440.0 Hz tuning fork is sounded together with an out-of-tune guitar string, and beat frequency of 3 Hz is heard. When the string is tightened, the frequency at which it vibrates increases, and the beat frequency is heard to decrease. What was the original frequency of the guitar string?

Activity 20: Ohm's Law
(Voltage-Current Sensor)

Preface

- *If* you are using the PASCO electronic Workbook specifically designed for this activity, then do the following:
1. Connect the *USB Link* to the computer's USB port.
2. Connect the *Voltage-Current Sensor* to the USB This will automatically launch the PASPORTAL window.
3. Choose the electronic Workbook entitled: **20 Ohm's Law WB.ds** and follow the directions in the Workbook.

Introduction

The purpose of this exploration is to study the relationship between voltage, current, and resistance for a simple resistor. This relationship is called Ohm's Law.

Use a Voltage-Current Sensor to measure the current through and voltage across a 10-ohm resistor and a 100-ohm resistor.

Learning Outcomes

You will be able to:

- Explore the basic concepts of Ohm's Law.

- Make predictions using Ohm's Law.

- Use a Voltage-Current Sensor to measure the voltage and current of a simple circuit.

- Compare your theoretical predictions to the experimental results.

Hypothesis

What is the relationship between current and voltage in a simple resistor?

Background

Ohm discovered that when the voltage (potential difference) across a resistor changes, the current through the resistor changes. He expressed this as "I=V/R" where I is current, V is voltage (potential difference), and R is resistance. Current is directly proportional to voltage and inversely proportional to resistance. In other words, as the voltage increases, so does the current.

For more information see Cutnell & Johnson, Physics, 6th ed., Vol. 1, Chapter 20, Section 20.2.

Materials

Equipment Needed	Qty	Equipment Needed	Qty
PASPORT Voltage-Current Sensor (PS-2115)	1	AC/DC Electronics Laboratory (EM-8656)	1
USB Link (PS-2100)	1	1.5 V D-Cell Battery (540-021)	2

Setup

Computer Setup

1. Plug the *USB Link* into the computer's USB port.

2. Plug the *PASPORT Voltage-Current Sensor* into the USB Link. This will automatically launch the PASPORTAL window.

3. Choose the appropriate DataStudio configuration file entitled **20 Ohm's Law CF.ds** and proceed with the following instructions.

SAFETY REMINDER	THINK SAFETY
• Follow the directions for using the equipment.	ACT SAFELY BE SAFE!

Equipment Setup

1. Follow the drawing to build the circuit.

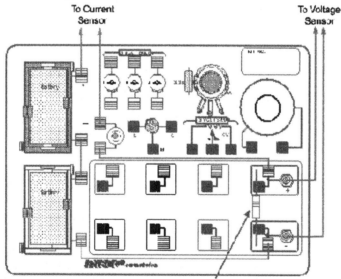

10 Ω RESISTOR (BROWN, BLACK, BLACK)

Part 1:

*   Place a 10-ohm resistor in the pair of component springs.

Part 2:

*   Place a 100-ohm resistor in the pair of component springs.

Note: The resistor color code for a 100-ohm resistor is brown, black, and brown.

Record Data

(Hint: Read this all the way through before you begin to take data.)

Record Data - 10 ohm Resistor

1.   Click **Start**. Hold down the button on the circuit board. Click **Stop**. Release the button.

2.   Record the current and voltage.

Current:	
Voltage:	

Record Data - 100 ohm Resistor

• Replace the 10-ohm resistor with a 100-ohm resistor.

3.   Click **Start**. Hold down the button on the circuit board. Click **Stop**. Release the button.

4.   Record the current and voltage.

Current:	
Voltage:	

•   Remember: The resistor color code for a 100-ohm resistor is brown, black, and brown.

Analyze

Observations

1.   What did you observe about the relationship between current and voltage?

2.   Based on your observations, is the relationship between current and voltage linear or inverse? Explain your reasoning.

Variables

1.   What was the independent variable in this activity (the one that you changed)?

2.   What was the dependent variable (the one that responded to the change)?

3.   How did the dependent variable respond?

Calculations

Use the values you recorded for the current and voltage to calculate the resistance using Ohm's Law, Equation 20.1, for the 10 ohm and 100 ohm resistors. $V = IR$

Resistance, 10 ohm (measured):
Resistance. 100 ohm (measured):

Equation 20.1

Synthesize

Verification

Compare the value you obtained with your data (measured) to the (theoretical) value. Calculate the percent difference for this experiment. Record your percent differences below.

Resistance (theoretical)	10 ohm	100 ohm
Percent Difference		

$$\% \text{ Difference} = \left|\frac{\text{measured - theoretical}}{\text{theorectical}}\right| \times 100\%$$

Error Analysis

What were the sources of error in this experiment?

Conclusions

What can you conclude about the relationship between current and voltage?

Do your results support your hypothesis?

Applications

What is the importance of Ohm's Law?

Extension Problem

The following problem is from Cutnell and Johnson, Physics, 6th ed., Volume One, Chapter 20, problem 3, page 617.

The filament of a light bulb has a resistance of 580 Ohms. A voltage of 120 V is connected across the filament. How much current is in the filament?

Activity 21: DC Series Wiring
(Voltage-Current Sensor)

Preface

> * *If* you are using the PASCO electronic Workbook specifically designed for this activity, then do the following:
> 1. Connect the *USB Link* to the computer's USB port.
> 2. Connect the *Voltage-Current Sensor* to the USB Link. This will automatically launch the PASPORTAL window.
> 3. Choose the electronic Workbook entitled: **21 DC Series Wiring WB.ds** and follow the directions in the Workbook.

Introduction

An array of resistors will have different measured resistances depending on how they are connected. If they are connected in series (end-to-end), will their total resistance equal the sum of all of their individual resistances, or will the total resistance be something different?

Use a Voltage-Current Sensor to measure current and voltage in a circuit consisting of a voltage source and several resistors.

Determine what happens to the current as more resistors are added in series to the circuit.

Learning Outcomes

You will be able to:

*   Explore the relationship between voltage, current, and resistance for resistors in series.

*   Measure the voltage across the voltage source and across the resistor and compare them.

*   Understand how resistors work in series.

*   Compare the theoretical values for resistors in series with the measured values.

Hypothesis

How will the current in a series circuit change as more resistors are added to the circuit? Assume that the voltage remains constant.

How will the voltage across the voltage source compare to the voltage across the resistors in a series circuit?

Background

There are many circuits in which more than one device is connected to a voltage source. We will introduce one method by which such connections may be made, namely, series wiring. Series wiring means that the devices are connected in such a way that there is the same electric current through each device. The figure below shows a circuit in which two different devices, represented by R_1 and R_2, are connected in series with a battery. Note that if the current in one resistor is interrupted, the current in the other is too.

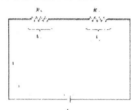

This could occur, for example, if two light bulbs were connected in series and the filament of one bulb broke. Because of the series wiring, the voltage V supplied by the battery is divided between the two resistors.

The drawing indicates that the portion of the voltage across R_1 is V_1, while the portion across R_2 is V_2, so $V = V_1 + V_2$. For each resistance, the definition of resistance is R = V/I, which may be solved for the corresponding voltage V = IR. Therefore, we have

$$V = V_1 + V_2 = IR_1 + IR_2 = I(R_1 + R_2) = IR_s$$

where R_s is called the equivalent resistance of the series circuit. Thus, two resistors in series are equivalent to a single resistor whose resistance is $R_s = R_1 + R_2$, in the sense that there is the same current through R_s as there is through the series combination of R_1 and R_2. This line of reasoning can be extended to any number of resistors in series, with the result that $R_s = R_1 + R_2 + R_3 + ...$

For more information see Cutnell & Johnson, Physics, 6th ed., Vol. 2, Chapter 20, Section 20.6.

Materials

Equipment Needed	Qty	Equipment Needed	Qty
PASPORT Voltage-Current Sensor (PS-2115)	1	1.5 V D-Cell Battery	2
USB Link (PS-2100)	1	Alligator Clips	2
AC/DC Electronics Laboratory (EM-8656)	1		

Setup

Computer Setup

1. Plug the *USB Link* into the computer's USB port.

2. Plug the *PASPORT Voltage-Current Sensor* into the USB Link. This will automatically launch the PASPORTAL window.

3. Choose the DataStudio configuration file entitled **21 DC Series Wiring CF.ds** and proceed with the following instructions.

SAFETY REMINDER	THINK SAFETY
• Follow the directions for using the equipment.	ACT SAFELY BE SAFE!

Equipment Setup

1.   Insert 2 D cell batteries into the AC/DC Electronics Laboratory.

2.   Put together the Electronics Lab and Voltage-Current Sensor as shown.

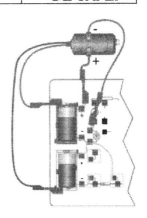

Note: The diagram is not to scale and it does not show the USB Link connected to the end of the Voltage-Current Sensor.

Record Data

(Hint: Read this all the way through before you begin to take data.)

Measure the Voltage of the Voltage Source

1.   Insert a 100-ohm resistor into the circuit (see the picture).

Note: The color code for a 100-ohm resistor is brown, black, and brown.

2.   Measure the voltage across the voltage source.

3.   Click **Start**. Hold down the button in the middle of the board. Click **Stop**. Release the button.

4.   Record the value for the voltage below.

Voltage (100 ohm):

5.   Calculate the theoretical value of the current using the measured voltage, the resistance of the resistor, and Ohm's Law. $V = IR$

Current (100 ohm)-theoretical:

Record Current and Voltage - 100 ohm Resistor

6.   Measure the voltage and the current.

7.   Click **Start**. Hold down the button in the middle of the board.

8.   Click **Stop**. Release the button. Record your measured current and voltage.

Current (100 ohm)-measured:	
Voltage (100 ohm):	

Record Voltage Across the 100 ohm Resistor

9.   Now measure the voltage across the resistor in the circuit.

10.   Remove the positive voltage alligator clip from the battery. Attach it to the right hand spring of the resistor.

11.   Click **Start**. Press the button in the middle of the circuit board. The voltage will appear in the display. Click **Stop**. Release the button. Record the voltage below.

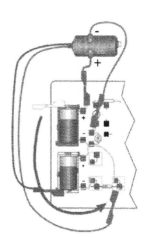

Voltage (100 ohm):

Observations

How did the voltage across the voltage source compare to the voltage across the resistor?

Measure the Voltage - Two 100 ohm Resistors

12.   Connect another 100-ohm resistor into the circuit (see picture).

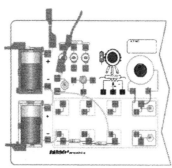

13.   Click **Start**. Hold down the button in the middle of the board. Click **Stop**. Release the button.

14.   Calculate the value of the current using the measured voltage, the resistance, and Ohm's Law.

Current (Two-100 ohm)-theoretical:

Record Current & Voltage - Two 100 ohm Resistors

15.   Click **Start**. Hold down the button in the middle of the board.

16.   Click **Stop**. Release the button. Record the values for the current and voltage below.

Current (Two - 100 ohm)-measured:	
Voltage (Two - 100 ohm)	

Observations

How does the current in the circuit with the two 100 ohm resistors compare to the current in the circuit with only one 100-ohm resistor?

If the current in the circuit with the two 100 ohm resistors is different than the current in the circuit with only one 100-ohm resistor, explain why.

Measure Voltage - Two 100 ohm and One Unknown

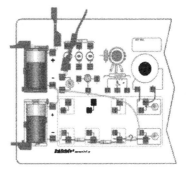

17.   Insert a resistor* of your choosing into the springs (see picture). Insert the value for the resistor below.

Unknown Resistor:

18.   Click **Start**. Hold down the button in the middle of the board. Click **Stop**. Release the button.

19.   Calculate the value of the current using Ohm's Law and the voltage displayed.

Current (Two-100 ohm + Unknown)-theoretical:

Record Current & Voltage – Two 100 ohm + Unknown

20.   Click **Start**. Hold down the button in the middle of the board.

21.   Click **Stop**. Release the button. Record the values for the current and voltage below.

Current (Two-100 ohm + Unknown)-measured:	
Voltage (Two-100 ohm + Unknown):	

Observations

When you added the resistor of your choice to the circuit and measured the current, how did the current compare to your previous measurements?

Synthesize

Verification

Fill in the table below and compare the value you obtained with your data (measured) to the (theoretical) value. Calculate the percent difference.

$$\% \text{ Difference} = \left| \frac{\text{measured - theoretical}}{\text{theoretical}} \right| \times 100\%$$

Resistor	Theoretical Current	Measured Current	% Difference
100 ohm			
Two 100 ohm			
Two 100 ohm + Unknown			

How did your calculated (theoretical) value for current through the resistors compare to the actual (measured) value for current?

Error Analysis

What were the sources of error in this experiment?

Conclusions

In general, what happens to the amount of current in a circuit as you add more resistors in series to the circuit? (Assume that the voltage remains constant.)

Do your results support your hypothesis?

Applications

How could you use a resistor that has a variable amount of resistance to control the amount of current in a circuit? Give an example of where this might be used.

Extension Problem

In Cutnell and Johnson, Physics, 6th ed., Volume Two, Chapter 20, problem 40, page 614.

The current in a 47 ohm resistor is 0.12 A. This resistor is in series with a 28-ohm resistor, and the series combination is connected across a battery. What is the battery voltage?

Activity 22: DC Parallel Wiring
(Voltage-Current Sensor)

Preface

> • *If* you are using the PASCO electronic Workbook specifically designed for this activity, then do the following:
> 1. Connect the *USB Link* to the computer's USB port.
> 2. Connect the *Voltage-Current Sensor* to the USB Link. This will automatically launch the PASPORTAL window.
> 3. Choose the electronic Workbook entitled: **22 DC Parallel Wiring WB.ds** and follow the directions in the Workbook.

Introduction

An array of resistors will have different measured resistances depending on how they are connected. If they are connected in parallel (side-by-side), will their total resistance equal the sum of all of their individual resistance, or will the total resistance be something different?

Use a Voltage-Current Sensor to measure current and voltage in a circuit consisting of a voltage source and several resistors.

Determine what happens to the current as more resistors are added in parallel to the circuit.

Learning Outcomes

You will be able to:

• Explore the relationship between voltage, current, and resistance for resistors in parallel.

• Measure the voltage across the voltage source and across the resistor and compare them.

• Understand how resistors work in parallel.

• Compare the theoretical values for resistors in series with the measured values.

Hypothesis

How will the current in a parallel circuit change as more resistors are added to the circuit? Assume that the voltage remains constant.

How will the voltage across the voltage source compare to the voltage across the resistors in a parallel circuit?

Background

Parallel wiring is another method of connecting electrical devices. Parallel wiring means that the devices are connected in such a way that the same voltage is applied across each device. The figure shows two resistors connected in parallel between the terminals of a battery. The picture is drawn so as to emphasize that the entire voltage of the battery is applied across each resistor. When two resistors R_1 and R_2 are connected as the figure shows, each receives current from the battery as if the other were not present. Therefore, R_1 and R_2 together draw more current from the battery than does either resistor alone. According to the definition of resistance, $R=V/I$, a larger current implies a smaller resistance. Thus, the two parallel resistors behave as a single equivalent resistance that is smaller than either R_1 or R_2.

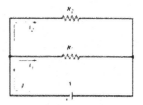

As in a series circuit, it is possible to replace a parallel combination of resistors with an equivalent resistor that results in the same total current and power for a given voltage as the original combination.

To determine the equivalent resistance for the two resistors in the figure on the previous page, note that the total current I from the battery is the sum of I_1 and I_2, where I_1 is the current in resistor R_1 and I_2 is the current in resistor R_2: $I = I_1 + I_2$. Since the same voltage V is applied across each resistor, the definition of resistance indicates that $I_1 = V/R_1$ and $I_2 = V/R_2$. Therefore,

$$I = I_1 + I_2 = \frac{V}{R_1} + \frac{V}{R_2} = V\left(\frac{1}{R_1} + \frac{1}{R_2}\right) = V\left(\frac{1}{R_p}\right)$$

where R_p is the equivalent resistance.

Hence, when two resistors are connected in parallel, they are equivalent to a single resistor whose resistance R_P can be obtained from $1/R_P = 1/R_1 + 1/R_2$. For any number of resistors in parallel, a similar line of reasoning reveals that

$$\frac{1}{R_p} = \frac{1}{R_1} + \frac{1}{R_2} + \frac{1}{R_3} + ...$$

For more information see Cutnell & Johnson, Physics, 6th ed., Vol. 2, Chapter 20, Section 20.7.

Materials

Equipment Needed	Qty	Equipment Needed	Qty
PASPORT Voltage-Current Sensor (PS-2115)	1	1.5 V D-Cell Battery	2
USB Link (PS-2100)	1	AC/DC Electronics Laboratory (EM-8656)	1

Setup

Computer Setup

1. Plug the *USB Link* into the computer's USB port.

2. Plug the *PASPORT Voltage-Current Sensor* into the USB Link. This will automatically launch the PASPORTAL window.

3. Choose the appropriate DataStudio configuration file entitled **22 DC Parallel Wiring CF.ds** and proceed with the following instructions.

SAFETY REMINDER	
• Follow the directions for using the equipment.	

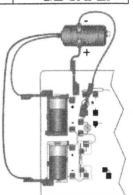

Equipment Setup

1. Insert 2 D cell batteries into the AC/DC Electronics Laboratory.

2. Put together the Electronics Lab and Voltage-Current Sensor as shown.

3. Insert a 100-ohm resistor into the circuit (see the picture).

The color code for a 100-ohm resistor is brown, black, and brown.

Note: The diagram is not to scale and it does not show the USB Link connected to the end of the Voltage-Current Sensor.

Record Data

(Hint: Read this all the way through before you begin to take data.)

Measure the Voltage of the Voltage Source

1. Measure the voltage across the voltage source.

2. Click **Start**. Hold down the button in the middle of the board. Click **Stop**. Release the button.

3. Record the value for the voltage below.

Voltage (100 ohm):

4. Calculate the theoretical value of the current using the measured voltage, the resistance of the resistor, and Ohm's Law. $V = IR$

Current (100 ohm)-theoretical:

Record Current and Voltage - 100 ohm Resistor

5.   Measure the voltage and the current.

6.   Click **Start**. Hold down the button in the middle of the board.

7.   Click **Stop**. Release the button. Record your measured current and voltage.

Current (100 ohm)-measured:	
Voltage (100 ohm):	

Record Voltage Across the 100 ohm Resistor

8.   Now measure the voltage across the resistor in the circuit.

9.   Remove the positive voltage alligator clip from the battery. Attach it to the right hand spring of the resistor.

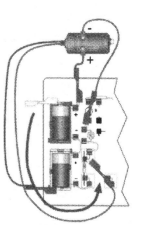

10. Click **Start**. Press the button in the middle of the circuit board. The voltage will appear in the display. Click **Stop**. Release the button. Record the voltage below.

Voltage (100 ohm):

Observations

How did the voltage across the voltage source compare to the voltage across the resistor?

Measure the Voltage - Two 100 ohm Resistors

11. Connect another 100-ohm resistor into the circuit (see picture).

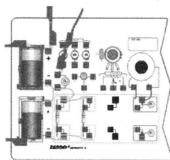

12. Click **Start**. Hold down the button in the middle of the board. Click **Stop**. Release the button.

13. Calculate the value of the current using the measured voltage, the resistance, and Ohm's Law.

Current (Two-100 ohm)-theoretical:

Record Current & Voltage - Two 100 ohm Resistors

14.   Click **Start**. Hold down the button in the middle of the board.

15.   Click **Stop**. Release the button. Record the values for the current and voltage below.

Current (Two - 100 ohm)-measured:	
Voltage (Two - 100 ohm)	

Observations

Is the current in the circuit with the two 100 ohm resistors different from the current in the circuit with only one 100 ohm resistor? If so, why?

Measure Voltage - Two 100 ohm and One Unknown

16.   Insert a resistor* of your choosing into the springs (see picture). Insert the value for the resistor below.

Unknown Resistor:

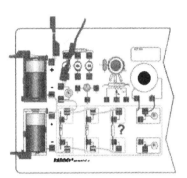

17.   Click **Start**. Hold down the button in the middle of the board. Click **Stop**. Release the button.

18.   Calculate the value of the current using Ohm's Law and the voltage displayed.

Current (Two-100 ohm + Unknown)-theoretical:

Record Current & Voltage – Two 100 ohm + Unknown

19.   Click **Start**. Hold down the button in the middle of the board.

20.   Click **Stop**. Release the button. Record the values for the current and voltage below.

Current (Two-100 ohm + Unknown)-measured:	
Voltage (Two-100 ohm + Unknown):	

Observations

When you added the resistor of your choice to the circuit and measured the current, how did the current compare to your previous measurements?

Synthesize

Verification

Fill in the table below and compare the value you obtained with your data (measured) to the (theoretical) value. Calculate the percent difference.

$$\% \text{ Difference} = \left| \frac{\text{measured - theoretical}}{\text{theorectical}} \right| \times 100\%$$

Resistor	Theoretical Current	Measured Current	% Difference
100 ohm			
Two 100 ohm			
Two 100 ohm + Unknown			

How did your calculated (theoretical) value for current through the resistors compare to the actual (measured) value for current?

Error Analysis

What were the sources of error in this experiment?

Conclusions

In general, what happens to the amount of current in a circuit as you add more resistors in parallel to the circuit? (Assume that the voltage remains constant.)

Do your results support your hypothesis?

Applications

If a circuit is wired in parallel, a component can be removed without breaking the circuit. For example, you could remove the middle resistor of the three and current would still flow through the circuit. Why are the electrical circuits in houses wired in parallel?

Extension Problem

The following problem is from Cutnell and Johnson, Physics, 6th ed., Volume Two, Chapter 20, problem 48, page 614.

What resistance must be placed in parallel with a 155-ohm resistor to make the equivalent resistance 115 ohms?

Activity 23: RC Circuit
(Voltage-Current Sensor)

Preface

Introduction

Have you ever shocked someone by dragging your feet on a carpet and then touching the person? As you drag your feet, you slowly accumulate charge. When you shock your friend, you rapidly let go of the charge (in other words, you discharge). For a short time, you are a capacitor!

A capacitor is one of the most commonly used electronic circuit components. They can store electric energy. Capacitors come in a variety of shapes and sizes, but they all behave in a similar way. Shown to the right is one type, known as a parallel-plate capacitor.

The voltage across a capacitor varies as it charges or discharges. Use a Voltage Sensor to measure the voltage across a capacitor as it charges and discharges in a resistor-capacitor circuit. Use your data to calculate the capacitance of your capacitor. (The value stamped on the capacitor can be up to 20% different than the actual capacitance.) Compare the calculated value of the capacitor to the stated value of the capacitor.

Learning Outcomes

You will be able to:

- Describe how the charge on a charging or discharging capacitor varies with time.

- State the correlation between the amount of charge on a capacitor and the voltage across it.

- Use technology to measure the voltage across the capacitor as it charges and discharges.

- Determine the time for the capacitor to charge to one-half of its maximum voltage.

- Calculate the capacitance based on the time to 'half-max'.

- Compare the measured capacitance of the capacitor to the stated value.

Hypothesis

How will the voltage across the capacitor change as the capacitor charges?

How will the voltage across the capacitor change as it discharges?

Background

Many electric circuits contain both resistors and capacitors. The diagram shows an example of a resistor-capacitor or RC circuit. Part (a) of the drawing shows the circuit at a time t after the switch has been closed and the battery has begun to charge up the capacitor plates. The charge on the plates builds up gradually to its equilibrium value of $q_o = CV_o$, where V_o is the voltage of the battery. Assuming that the capacitor is uncharged at time t = 0 s when the switch is closed, the capacitor charges exponentially as shown in Equation 23.1.

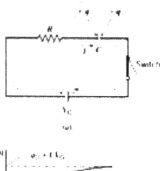

$$q = q_o \left(1 - e^{-t/RC}\right)$$ **Equation 23.1**

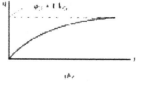

The exponential e has the value of 2.718, q is the amount of charge at any time, q_o is the maximum charge achieved, t is the amount of time elapsed, R is the resistance of the circuit and C is the value of the capacitor. The term RC in the exponent in Equation 23.1 is called the time constant τ of the circuit: $\tau = RC$. The time constant is the amount of time required for the capacitor to accumulate 63.2 percent of its equilibrium charge. The charge approaches its equilibrium value rapidly when the time constant is small and slowly when the time constant is large.

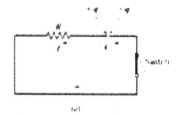

The diagram shows a circuit at a time t after a switch is closed to allow a charged capacitor to begin discharging. There is no battery in this circuit, so the charge +q on the left plate of the capacitor can flow counterclockwise through the resistor and neutralize the charge -q on the right plate. The graph in part b of the drawing shows that the charge begins at q_o when t = 0 s and decreases gradually toward zero. Smaller values of the time constant RC lead to a more rapid discharge. The time constant (RC) is also the amount of time required for a charged capacitor to lose 63.2 percent of its charge.

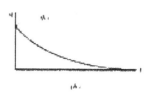

The time it takes to charge (or discharge) the capacitor to half full is called the half-life and is related to the capacitance and the resistance by Equation 23.2.

$$t_{1/2} = RC \ln 2$$

$$t_{1/2} = \tau \ln 2$$ **Equation 23.2**

For more information see Cutnell & Johnson, <u>Physics</u>, 6th ed., Volume Two, Chapter 20, Section 20.13.

Materials

Equipment Needed	Qty	Equipment Needed	Qty
PASPORTVoltage-Current Sensor (PS-2115)	1	AC/DC Electronics Laboratory (EM-8656)	1
USB Link (PS-2100)	1	1.5 V D-Cell Battery	2

Setup

Computer Setup

1. Plug the *USB Link* into the computer's USB port.

2. Plug the *PASPORT Voltage-Current Sensor* into the USB Link. This will automatically launch the PASPORTAL window.

3. Choose the appropriate DataStudio configuration file entitled **23 RC Circuit CF.ds** and proceed with the following instructions.

SAFETY REMINDER	THINK SAFETY
• Follow the directions for using the equipment.	THINK SAFETY ACT SAFELY BE SAFE!

Equipment Setup

1. Place a 100 K-ohm resistor (brown, black, yellow) in the pair of component springs nearest to the lower banana jack at the lower right corner of the AC/DC Electronics Lab Board.

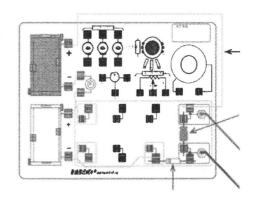

2. Connect a 100 microfarad (μF) capacitor between the component spring on the left end of the 100 K-ohm resistor and the component spring closest to the top banana jack.

3. Connect the banana plug patch cords from the Voltage-Current Sensor to the banana jacks on the AC/DC Electronics Lab Board.

4. Use jumper wires to connect the following component springs:

• From the negative end of the lower battery to the left end of the resistor.

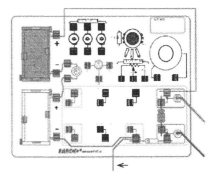

• From the positive end of the upper battery to the upper end of the capacitor.

• From the negative end of the upper battery to one of the component springs attached to the push-button switch.

• From the positive end of the lower battery to the other component spring attached to the push-button switch.

5. Connect one end of a jumper wire to the component spring at the left end of the resistor. Leave the other end of this wire unconnected.

6. For data recording during charging, leave the loose jumper wire disconnected. (On the left.)

7. For data recording during discharging, connect the end of the wire to the component spring above the capacitor. (On the right.)

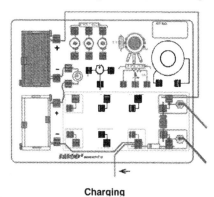

Charging

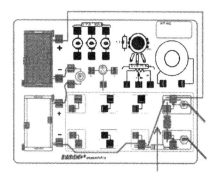

Discharging

Record Data

(Hint: Read this all the way through before you begin to take data.)

1. Fully discharge the capacitor by connecting a jumper wire between the component springs on either side of the capacitor for at least one second. Remove the jumper wire.

2. Click **Start**, then push and hold the red button until the voltage increases to a maximum. This will take more than 30 seconds.

3. At this point, release the red button and connect the loose wire to the component spring connected to the upper end of the capacitor. Do this as quickly as possible.

•   Note: The voltage should decrease until it reaches 0 volts. DataStudio will automatically stop taking data at 105 seconds.

Analyze

Observations

1. When the circuit was charging, how did the voltage change with time? Did it increase at a steady rate? (i.e. was the voltage vs. time graph a straight line?)

2. When the circuit was discharging, how did the voltage change with time? Did it decrease at a steady rate? (i.e. was the voltage vs. time graph a straight line?)

Data Analysis

In order to calculate the capacitance, you will need to determine the time it took your circuit to reach half of the total voltage. (This time is known as the half-life.) The next pages describe how to find the half-life from the voltage vs. time plot. Use this procedure to determine the half-life of your own data.

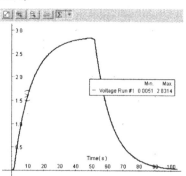

- Click the **Statistics** button. The Legend box shows the maximum voltage. (In this example, the "Maximum Voltage" is 2.8314 V.) Divide the maximum voltage by two. This will be your "Half-maximum Voltage".

• Use the **Zoom Select** tool to expand the increasing part of the graph. • Click the **Smart Tool** and move the **Smart Tool** along the expanded voltage vs. time curve until the y-value (voltage) is equal to the "Half-max Voltage" value you calculated. • The time it takes to reach "half-max" voltage from the minimum (zero) is the half-life. In the example, the "Time to Half-Max, Charging" is 8.64 seconds.	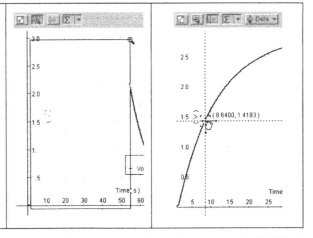

- For the discharging part, determine how much time it took to reach "half-max" from the time the circuit began to discharge. First, find the "Time of Maximum Voltage".

- Use the **Zoom Select** tool to expand the discharge portion of the graph.

- Use the **Smart Tool** to find the time that the discharge begins, or the "Time of Maximum Voltage, Discharging". (In the example, the "Time of Maximum Voltage, Discharging" occurs at 51.38 seconds.)

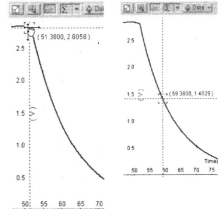

- Move the **Smart Tool** along the voltage vs. time curve until the y-value (voltage) is equal to the "Half-maximum Voltage" value.

- The time to reach the "half-max" voltage from its maximum voltage is the half-life. In the example, the "Time of Half-Max Voltage, Discharging" occurs at 59.38 seconds.

Item	Value
Voltage, Maximum (V)	
Voltage, Half-maximum (V)	
Time, Half-max. Voltage, Charging (s)	
Time, Maximum Voltage (s)	
Time, Half-max. Voltage, Discharging (s)	

Calculations

1. To determine the half-life from the discharging data, subtract the "Time, Maximum Voltage, Discharging" from the "Time, Half-Max Voltage, Discharging". Record the time as "Time to Half-Max, Discharging".

2. Average the two values of Time to Half-Max. Record as "Time to Half-Max, Average".

3. Use the average value for the half-life and Equation 23.2 to calculate the capacitance of your circuit. Record the calculated capacitance.

$$C = \frac{t_{1/2}}{R(\ln 2)}$$

105 microfarads

4. Record the accepted value (on the side of the capacitor) for the capacitance.

100 microfarads

5. Calculate and record a percent difference between the calculated and accepted values for capacitance.

$$\% \text{ Difference} = \left| \frac{\text{measured - theoretical}}{\text{theorectical}} \right| \times 100\%$$

Note: Capacitors may vary by 20% from the accepted value.

Synthesize

Error Analysis

What were the sources of error in this experiment?

Conclusions

Do your results support your hypothesis?

Applications

When you use the flash on a camera, it often takes a few seconds before you can use the flash again. Why is this?

Extension Problem

The following problem is from Cutnell and Johnson, Physics, 6th ed., Volume Two, Chapter 20, problem 96, page 616.

An electronic flash attachment for a camera produces a flash by using the energy stored in a 750-microfarad capacitor. Between flashes, the capacitor recharges through a resistor whose resistance is chosen so the capacitor recharges with a time constant of 3.0s. Determine the value of the resistance.

Activity 24: Magnetic Field Around a Wire
(Voltage-Current Sensor, Magnetic Field Sensor)

Preface

- *If* you are using the PASCO electronic Workbook specifically designed for this activity, then do the following:
1. Connect the *USB Links* to the computer's USB ports.
2. Connect the *Voltage-Current Sensor* to one of the USB Links and the *Magnetic Field Sensor* to the other. This will automatically launch the PASPORTAL window.
3. Choose the electronic Workbook entitled: **24 Mag Field around Wire WB.ds** and follow the directions in the Workbook.

Introduction

A current carrying wire experiences a magnetic force when placed in a magnetic field that is produced by an external source, such as a permanent magnet. A current- carrying wire also produces a magnetic field of its own. Hans Christian Oersted (1777-1851) first discovered this effect in 1820 when he observed that a current-carrying wire influenced the orientation of a nearby compass needle. The compass needle aligns itself with the net magnetic field produced by the current and the magnetic field of the earth. Oersted's discovery, which linked the motion of electric charges with the creation of a magnetic field, marked the beginning of an important discipline called electromagnetism.

Use a Magnetic Field Sensor and Voltage-Current Sensor to find the magnetic field strength and current in a loop of wire.

Graph the magnetic field strength versus the current to help you find the permeability of free space.

Compare your value to the accepted value.

Learning Outcomes

You will be able to:

- Gain some understanding of magnetic fields around a current-carrying wire.

- Determine how magnetic field strength relates to current.

- Find the permeability of free space and compare it to the accepted value.

Hypothesis

What will happen to the strength of the magnetic field around a wire as the current is changed?

As the number of coils increases and the current remains constant, what happens to the magnetic field?

Background

Experimentally, it is found that the magnitude B of the magnetic field produced by a long, straight wire is directly proportional to the current I and inversely proportional to the radial distance r from the wire as shown in Equation 24.1.

$$B \propto \frac{I}{r}$$
Equation 24.1

As usual, this proportionality is converted into an equation by introducing a proportionality constant, which, in this instance, is written as shown in Term 24.1.

$$\frac{\mu_o}{2\pi}$$
Term 24.1

Thus, the magnitude of the magnetic field around a long, straight wire is given by Equation 24.2.

$$B = \frac{\mu_o I}{2\pi r}$$

The constant "μ_o" is known as the permeability of free space, and its value is shown in Equation 24.3.

Equation 24.2

$$\mu_o = 4\pi \times 10^{-7} T \cdot \frac{m}{A}$$

If a current-carrying wire is bent into circular loop, the magnetic field lines around the loop have the pattern shown:

Equation 24.3

If a current-carrying wire is bent into circular loop, the magnetic field lines around the loop have the pattern shown:

At the center of a loop of radius R, the magnetic field is perpendicular to the plane of the loop and has the value shown in Equation 24.4 where I equals the current in the loop.

Often, the loop consists of N turns of wire that are wound sufficiently close together so that they form a flat coil with a single loop. In this case, the magnetic fields of the individual turns add together to give a net field that is N times greater than that of a single loop. For such a coil the magnetic field at the center is given by Equation 24.5.

$$B = \frac{\mu_o I}{2R}$$
Equation 24.4

For more information see Cutnell & Johnson, Physics, 6th ed., Volume Two, Chapter 21, Section 21.7.

$$B = N \frac{\mu_o I}{2R}$$
Equation 24.5

Materials

Equipment Needed	Qty	Equipment Needed	Qty
PASPORT Voltage-Current Sensor (PS-2115)	1	Power Supply, 18 V DC, 5 A (SE-9720)	1
PASPORT Magnetic Field Sensor (PS-2112)	1	Large Rod Base (ME-8735)	1
USB Link (PS-2100)	2	Patch Cords, 4 mm, Red (5) (SE-9750)	1
Clamp, Buret (SE-9446)	1	Wire*	1
Tape	1 in.	Support Rod, 45 cm (ME-8736)	1

Setup

Computer Setup

1. Plug the *USB Links* into the computer's USB ports.

2. Plug the *PASPORT Sensors* into the USB Links. This will automatically launch the PASPORTAL window.

3. Choose the DataStudio configuration file entitled **24 Mag Field around Wire CF.ds** and proceed with the following instructions.

SAFETY REMINDER	THINK SAFETY
• Follow the directions for using the equipment.	**THINK SAFETY** **ACT SAFELY** **BE SAFE!**

Equipment Setup

1. Create a loop of wire by wrapping the wire around an object of reasonable diameter. Make sure you count the number of times you wrap the wire. Enter this value below. Also measure and record the coil's diameter.

Note: Try to use an even number of turns to make your calculations easier. (ex. 10, 50, 100).

Number of wire turns (N):	
Diameter of coil meters (2R):	

2. Wrap some tape around the coil to hold it together. Once you have created a coil of wire, sand the wire ends to create a better contact area.

 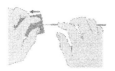

3. Attach the utility clamp to the stand. Open the clamp so the Magnetic Field Sensor can fit inside. Attach the sensor so the probe points down. Place the coil under the sensor.

Note: The Magnetic Field Sensor should be in the middle of the coil. The end of the sensor should be flush against the table.

4. Set up the circuit as shown in the diagram to measure the current through your coil.

Note: Make sure the current on the power supply is initially set to zero.

Record Data

(Hint: Read this all the way through before you begin to take data.)

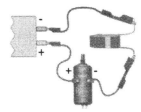

1. Click **Start**. Starting at zero amps on the power supply, slowly and smoothly increase the current.

2. Click **Stop** when you reach 1 ampere.

Analyze

Observations

What happens to the magnetic field when the current increases? Explain.

Data Analysis

To determine what happens, you need to find the slope of the graph.

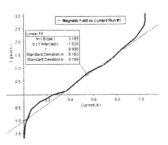

• Click the **Scale to Fit** button if needed. Highlight the flat-middle part of the graph. Click the **Fit** button. Select **Linear Fit** from the menu. The Legend box shows the slope ("m"). Record the slope.

Variables

What were the variables in this activity?

Which of those did you control?

Which did you manipulate and how did it respond?

Calculations

To find the value for the permeability of free space, first convert your slope from gauss units to tesla units by using the following conversion (1 gauss = 10^{-4} tesla).

Slope (gauss/A)	
Slope (tesla/A)	

Rearrange the equation and use it to solve for the permeability of free space (μ_o).
$$B = N\frac{\mu_o I}{2R}$$

Permeability of free space (measured):

Note: The slope is B/I, the ratio of magnetic field to current.

Synthesize

Verification

Compute percent difference to compare your value for permeability of free space to the accepted value. Record your difference below.

$$\% \text{ Difference} = \left|\frac{\text{measured - theoretical}}{\text{theorectical}}\right| \times 100\%$$

Permeability of free space (theoretical): 1.26×10^{-6} T m/A

Percent Difference:

Was your value to close the accepted value? Why or why not?

Error Analysis

What were the sources of error in this experiment?

Conclusions

How does the number of turns affect the strength of the magnetic field?

Do your results support your hypothesis?

Applications

If the current remains constant, what do you expect to happen to the magnetic field for an infinite number of loops?

Extension Problem

The following problem is from Cutnell and Johnson, Physics, 6th ed., Volume Two, Chapter 21, problem 48, page 656.

What must be the radius of a circular loop of wire so the magnetic field at its center is 1.8 x 10-4 T when the loop carries a current of 12 A?

Activity 25: Magnetic Field of a Solenoid
(Voltage-Current Sensor, Magnetic Field Sensor)

Preface

- *If* you are using the PASCO electronic Workbook specifically designed for this activity, then do the following:
1. Connect the *USB Links* to the computer's USB ports.
2. Connect the *Voltage-Current Sensor* to one of the USB Links and the *Magnetic Field Sensor* to the other. This will automatically launch the PASPORTAL window.
3. Choose the electronic Workbook entitled: **25 Mag Field of Solenoid WB.ds** and follow the directions in the Workbook.

Introduction

Discover what the magnetic field is like inside a coil of wire known as a solenoid.

Calculate the magnetic field strength inside a solenoid.

Use a Magnetic Field Sensor to measure the magnetic field strength inside a solenoid.

Examine the relationship of the magnetic field strength to the position inside of a solenoid.

Learning Outcomes

You will be able to:

- Measure the axial component of the magnetic field inside the solenoid.

- Calculate a theoretical value for the magnetic field of the solenoid.

- Compare the experimental and theoretical values of the magnetic field.

Hypothesis

How does the strength of the magnetic field inside a solenoid relate to the position inside?

Is the magnetic field the same strength at every location within the solenoid, or is it different from one location to another?

Background

A solenoid is a long coil of wire in the shape of a helix (see Figure below). If the wire is wound so the turns are packed close to each other and the solenoid is long compared to its diameter, the magnetic field lines have the appearance shown in the drawing.

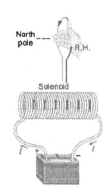

Notice that the field inside the solenoid and away from its ends is nearly constant in magnitude and directed parallel to the axis. The Right Hand Rule 2 (RHR-2) gives the direction of the field inside the solenoid, just as it is for a circular current loop.

The magnitude of the magnetic field in the interior of a long solenoid is shown in Equation 25.1 where n is the number of turns per unit length of the solenoid, μ_o is the permeability of free space, and I is the current.

$$B = \mu_0 n I$$

If, for example, the solenoid contains 100 turns and has a length of 0.05m, the number of turns per unit length is n = (100 turns)/(0.05m) = 2000 turns/m. The magnetic field outside the solenoid is not constant and is much weaker than the interior field. In fact, the magnetic field outside is nearly zero if the length of the solenoid is much greater than its diameter.

Equation 25.1

$$\mu_o = 4\pi \times 10^{-7} \frac{T \cdot m}{A}$$

For more information see Cutnell & Johnson, Physics, 6th ed., Vol. 2, Chapter 21, Section 21.7.

Materials

Equipment Needed	Qty	Equipment Needed	Qty
PASPORT Voltage-Current Sensor (PS-2115)	1	Power Supply, 18 V DC, 5 A (SE-9720)	1
PASPORT Magnetic Field Sensor (PS-2112)	1	Primary and Secondary Coil (SE-8653)	1
USB Link (PS-2100)	2	Meter Stick	1

Setup

Computer Setup

1. Plug the *USB Links* into the computer's USB ports.

2. Plug the *PASPORT Sensors* into the USB Links. This will automatically launch the PASPORTAL window.

3. Choose the appropriate DataStudio configuration file entitled **25 Mag Field of Solenoid CF.ds** and proceed with the following instructions.

SAFETY REMINDER • Follow the directions for using the equipment.	

Equipment Setup

1. Use only the outer coil of the Primary/Secondary Coil set. Measure the length, in meters, of the solenoid coil. Insert the number of turns for the coil. (Your instructor will provide you with this value.). Record the length and number of turns below.

Note: When measuring the coil, make sure that you only measure the length of the solenoid with the wrapped coil and not the entire solenoid.

Length of solenoid (m):

Number of turns for the coil (n):

Note: The PASCO Model SE-8653 Primary/Secondary Coils has 2920 turns on the secondary coil.

2. Use patch cords to connect the output of the DC Power Supply to the input jacks on the solenoid.

3. Using patch cords attach the current portion of the Voltage-Current Sensor to the solenoid.

4. Position the solenoid and Magnetic Field Sensor so the end of the sensor can be placed inside the solenoid.

Record Data

(Hint: Read this all the way through before you begin to take data.)

1. Turn on the DC Power Supply and set it to 10 V.

2. Insert the Magnetic Field Sensor to the center of the solenoid.

3. Click **Start**.

4. Record the current and magnetic field strength in the table below.

Current (A):	
Magnetic Field (measured):	

5. Click **Stop**.

6. Plot the field strength as you slowly insert the Magnetic Field Sensor into the solenoid.

7. Start with the Magnetic Field Sensor just on the outside of the Solenoid. Click **Start**.

8. Click the **Keep** button. Your first value for the Magnetic Field will appear in the first cell.

9. Move the Magnetic Field Sensor 1 cm at a time. Click the **Keep** button each time you move the sensor into the solenoid by one centimeter.

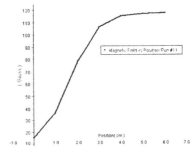

10. Collect 7 values. Click **Stop** when you are finished.

Analyze

Data Analysis

• Click the **Scale to Fit** button.

Observations

Is the graph of your data what you expected to find? Why?

Is the magnetic field inside the solenoid variable or linear? Why?

Variables

What was the independent variable in this activity (what did you change)?

What was the dependent variable (what did you measure)?

How did the quantity that you measured respond to the independent variable?

Calculations

Calculate the theoretical value for the magnetic field inside the solenoid using the current you found above, the length of the coil, the number of turns in the coil, and the permeability of free space (μ_o).

Magnetic Field (theoretical):

$$B = \frac{\mu_o n I}{L}$$

Synthesize

Verification

Compute the percent difference of the measured value to the theoretical value for the Magnetic Field. Record your percent difference below.

$$\% \text{ Difference} = \left| \frac{\text{measured - theoretical}}{\text{theorectical}} \right| \text{x}100\%$$

Percent Difference:

Error Analysis

What were the sources of error in this experiment?

Conclusions

Do your results support your hypothesis?

Applications

Solenoids are an important aspect of automated controls. Solenoids are used in common household appliances. Can you name a few? (Hint: Start with the washing machine.)

Extension Problem

A long solenoid consists of 1400 turns of wire and has a length of 0.65 m. There is a current of 4.7 A in the wire. What is the magnitude of the magnetic field within the solenoid?

Activity 26: Faraday's Law
(Voltage-Current Sensor)

Preface

> - *If* you are using the PASCO electronic Workbook specifically designed for this activity, then do the following:
> 1. Connect the *USB Link* to the computer's USB port.
> 2. Connect the *Voltage-Current Sensor* to the USB Link. This will automatically launch the PASPORTAL window.
> 3. Choose the electronic Workbook entitled: **26 Faraday's Law WB.ds** and follow the directions in the Workbook.

Introduction

When electricity is passed through a conducting wire a magnetic field can be detected near the wire. Michael Faraday was one of the first scientists to reverse the process. The purpose of this activity is to measure the electromotive force (emf) induced in a coil by a magnet dropping through the center of a coil.

Use the Voltage-Current Sensor to measure the voltage induced in a coil as a bar magnet moves through the coil.

Learning Outcomes

You will be able to:

- Measure the voltage induced in a coil by a magnet that moves through the coil.

- Determine the area under the voltage versus time curve.

- Compare the 'flux' (voltage x time) induced by one end of the magnet to the 'flux' induced by the other end.

- State whether or not the activity confirms Faraday's statement about voltage induced by a magnet.

Hypothesis

You can send electricity through a conducting wire to make a magnetic field. Is the reverse possible?

Can you use a magnet and a conducting wire to make electricity?

Background

When a magnet is passed through a coil there is a changing magnetic flux through the coil that induces an electromotive force (emf) in the coil. According to Faraday's Law of Induction shown in Equation 26.1, the emf, epsilon, depends on the number of coils, N, and the rate of change of flux through the coils (see Term 26.1).

In this activity, a plot of the EMF versus time is made and the area under the curve is found by integration. This area represents the flux as shown in Equation 26.2.

$$\varepsilon = -N\frac{\Delta\phi}{\Delta t}$$

Equation 26.1

$$\frac{\Delta\phi}{\Delta t}$$

Term 26.1

$$\varepsilon\Delta t = -N\Delta\phi$$

Equation 26.2

For more information see Cutnell & Johnson, <u>Physics</u>, 6th ed., Vol. 2, Chapter 22, Section 22.4.

Materials

Equipment Needed	Qty	Equipment Needed	Qty
PASPORT Voltage-Current Sensor (PS-2115)	1	AC/DC Electronics Laboratory (EM-8656)	1
USB Link (PS-2100)	1	Bar Magnet (EM-8620)	1

Setup

Computer Setup

1. Plug the *USB Link* into the computer's USB port.

2. Plug the *PASPORT Voltage-Current Sensor* into the USB Link. This will automatically launch the PASPORTAL window.

3. Choose the appropriate DataStudio configuration file entitled **26 Faraday's Law CF.ds** and proceed with the following instructions.

SAFETY REMINDER	THINK SAFETY
• Follow the directions for using the equipment.	ACT SAFELY BE SAFE!

Equipment Setup

1. Put alligator clips on the ends of the Voltage-Current Sensor leads.

2. Attach a Voltage-Current Sensor lead to one component spring next to the inductor coil on the circuit board. Attach the other lead to the other component spring next to the coil.

3. Arrange the circuit board so the corner with the coil is beyond the edge of the table, and a magnet dropped through the coil can fall freely.

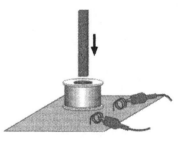

Note: The bar magnet will be dropped through the coil. Make sure that the magnet does not strike the floor, or it may break.

Record Data

(Hint: Read this all the way through before you begin to take data.)

1.   Hold the magnet so that the south end is about 2 cm above the coil.

Note: If you are using the PASCO Model EM-8620 Alnico Bar Magnet the North end is indicated by the narrow groove near one end. For the first run, hold the magnet with groove end 'up'.

2.   Click **Start**. Let the magnet drop through the coil.

3.   Data recording will begin when the magnet falls through the coil and the voltage from the coil reaches 0.05 V. Data recording will end automatically after 0.5 seconds.

Analyze

Data Analysis

- Click the **Scale to Fit** button to rescale the graph if needed.

- Use the cursor to select a region around the first peak of the voltage plot. The value for 'Area' appears in the legend in the graph. Record the Area below.

- Perform the same procedure for the second peak and record its Area below.

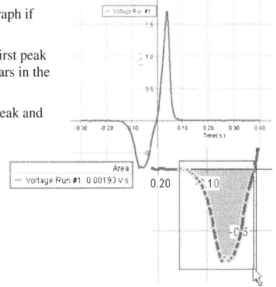

Peak	Area
1	
2	

Observations

1.   Is the incoming flux equal to the outgoing flux?

2.   Why are the peaks opposite in direction?

Variables

1.   What were the variables in this activity?

2.   Which of those did you control?

3.   Which did you manipulate and what was the response?

Comparison

1.   How do the two values of area compare to each other?

2.   How close is the sum to 'zero'?

Note: According to Faraday, the two areas should be zero when added together.

Synthesize

Error Analysis

What were the sources of error in this experiment?

Conclusions

If the two peaks are supposed to be equal but opposite, why aren't they in this case?

Do your results support your hypothesis?

Applications

Where do we see Faraday's Law of Induction being used in your home?

Extension Problem

The following problem is from Cutnell and Johnson, Physics, 6th ed., Volume Two, Chapter 22, problem 18, page 690.

In each of two coils the rate of change of the magnetic flux in a single loop is the same. The emf induced in coil 1, which has 184 loops, is 2.82 V. The emf induced in coil 2 is 4.23 V. How many loops does coil 2 have?

Activity 27: Polarization
(Light Sensor, Rotary Motion Sensor)

Preface

> • *If* you are using the PASCO electronic Workbook specifically designed for this activity, then do the following:
> 1. Connect the *USB Links* to the computer's USB ports.
> 2. Connect the *Light Sensor* to one of the USB Links and the *Rotary Motion Sensor* to the other. This will automatically launch the PASPORTAL window.
> 3. Choose the electronic Workbook entitled: **27 Polarization WB.ds** and follow the directions in the Workbook.

Introduction

The purpose of this activity is to determine the relationship between the intensity of the transmitted light through two polarizers and the angle, theta, between the axes of the two polarizers. You will also verify Malus' Law.

Use the Light Sensor to measure the relative intensity of light that passes through two polarizers as you change the angle of the second polarizer relative to the first polarizer. Use the Rotary Motion Sensor to measure the angle of the second polarizer relative to the first polarizer.

Learning Outcomes

You will be able to:

Determine the relationship between light intensity and angle for polarized light.

Explore the effects of polarization on transmitted light.

Confirm Malus' Law by measuring the intensity of transmitted light through polarizers as the angle between the polarizers is changed.

Hypothesis

If the intensity of transmitted light is at its maximum when the axes of the polarizers are parallel, and at its minimum when the axes are perpendicular, how does the intensity depend on the angle between the axes for angles between 0 and 90 degrees?

Background

One of the essential features of electromagnetic waves is that they are transverse waves, and because of this feature they can be polarized. A polarizer only transmits the component of light that is vibrating in a particular plane to pass through it. This plane forms the "axis" of polarization. Unpolarized light vibrates in all planes perpendicular to the direction of propagation.

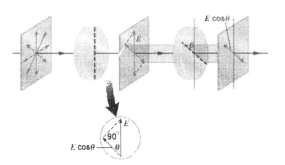

Two polarizers may be used to adjust the direction and intensity of the light. Changing the angle theta between the transmission axes of the polarizer and analyzer can do this.

Since the intensity of the light varies as the square of the electric field, the light intensity transmitted through the second filter is given by Equation 27.1 where So is the intensity of the light passing through the first filter and *theta* is the angle between the polarization axes of the two filters.

$$\bar{S} = \bar{S}_0 \cos^2 \theta$$

Equation 27.1

Consider the two extreme cases illustrated by this equation:

If theta is zero, the second polarizer is aligned with the first polarizer, and the value of $\cos^2(\text{theta})$ is one. Thus the intensity transmitted by the second filter is equal to the light intensity that passes through the first filter. This case will allow maximum intensity to pass through.

If theta is 90°, the second polarizer is oriented perpendicular to the plane of polarization of the first filter, and the value $\cos^2(90°)$ gives zero. Thus no light is transmitted through the second filter. This case will allow minimum intensity to pass through.

For more information see Cutnell & Johnson, Physics, 6th ed., Vol. 2, Chapter 24, Section 24.6.

Materials

Equipment Needed	Qty	Equipment Needed	Qty
PASPORT Light Sensor (PS-2106)	1	Basic Optics System (OS-8515)	1
PASPORT Rotary Motion Sensor (PS-2120)	1	Aperture Bracket (OS-8534)	1
USB Link (PS-2100)	2	Polarization Analyzer (OS-8533)	1

Setup

Computer Setup

1. Plug the *USB Links* into the computer's USB ports.

2. Plug the *PASPORT Sensors* into the USB Links. This will automatically launch the PASPORTAL window.

3. Choose the DataStudio configuration file entitled **27 Polarization CF.ds** and proceed with the following instructions.

SAFETY REMINDER	
• Follow the directions for using the equipment.	**THINK SAFETY ACT SAFELY BE SAFE!**

Equipment Setup

1. Attach the Rotary Motion Sensor to the Polarization Analyzer.

2. Attach the plastic band to the Polarization Analyzer and Rotary Motion Sensor.

3. Place the Basic Optics Light Source and a polarizer on the Optics Bench.

4. Turn the polarizer so the zero degree mark is next to the reference peg on the accessory holder.

5. Mount the Polarization Analyzer on the Optics Bench. Turn on the Light Source.

6. Turn the Polarization Analyzer so the transmitted light through the polarizers is minimized.

7. Mount the Light Sensor on the Aperture Bracket and attach the Aperture Bracket to the Aperture Bracket Holder. Put the Aperture Bracket Holder onto the Optics Bench.

8. Rotate the Aperture Disk so the open circular aperture is in line with the opening to the Light Sensor. Move the Light Source and the Light Sensor so they are as close as possible to the polarizers.

Note that the Aperture Bracket must not interfere with the movement of the Rotary Motion Sensor.

Note: It is very important the polarizers be at the minimum setting (that is, so that they transmit the minimum amount of light.)

9. You will rotate the Polarization Analyzer as you record data. You will rotate the polarizer through one complete turn (360 degrees).

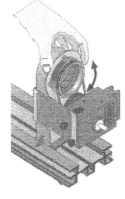

10. Remember to turn off the light source after you have completed taking your data.

Record Data

(Hint: Read this all the way through before you begin to take data.)

1. Click **Start**. Slowly rotate the Polarization Analyzer so the angle increases.

2. Click **Stop** when you have rotated the Polarization Analyzer through at least 360 degrees.

Analyze

Data Analysis

This is an example of what your graph should look like.

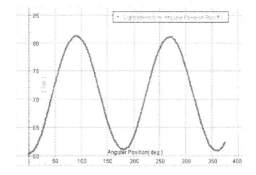

- This graph plots the light intensity versus the cosine squared of the angle. The coefficient of linear regression, r, is a measure of how well the data fit a straight line.

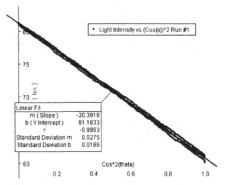

1. Click the **Fit** button. Select **Linear Fit** from the menu.

- The smaller the value of Standard Deviation m, the better the fit between the data and a straight line.

Observations

1. What is the shape of the graph of Light Intensity versus Angular Position?

2. Why is the graph of the Light Intensity vs. Cosine Squared linear?

Variables

What is the independent variable in this activity (what did you vary)?

What changed as a result?

How did that variable change relative to the independent variable?

Synthesize

Error Analysis

What were the sources of error in this experiment?

Conclusions

What can you conclude about the relationship between light intensity and the angle of the polarizers?

Do your results support your hypothesis?

Applications

What are some products that use polarized lenses?

Extension Problem

The following problem is from Cutnell and Johnson, Physics, 6th ed., Volume Two, Chapter 24, problem 34, page 747.

Unpolarized light whose intensity is 1.10 W/m² is incident on the polarizer. (a) What is the intensity of the light leaving the polarizer? (b) if the analyzer is set at an angle of theta = 75 degrees with respect to the polarizer, what is the intensity of the light that reaches the photocell?

Activity 28: Diffraction of Light
(Light Sensor, Rotary Motion Sensor)

Preface

- *If* you are using the PASCO electronic Workbook specifically designed for this activity, then do the following:
1. Connect the *USB Links* to the computer's USB ports.
2. Connect the *Light Sensor* to one of the USB Links and the *Rotary Motion Sensor* to the other. This will automatically launch the PASPORTAL window.
3. Choose the electronic Workbook entitled: **28 Diffraction of Light WB.ds** and follow the directions in the Workbook.

Introduction

The wave nature of light can be investigated by studying diffraction patterns.

Use the Light Sensor to measure the intensity of the maxima in a double-slit diffraction pattern created by monochromatic laser light passing through a double-slit. Use the Rotary Motion Sensor (RMS) to measure the relative positions of the maxima in the diffraction pattern.

Use the Light Sensor to measure the intensity of the maxima in a single-slit diffraction pattern created by monochromatic laser light passing through a single-slit.

Use the Rotary Motion Sensor (RMS) to measure the relative positions of the maxima in the diffraction pattern.

Record and display the light intensity and the relative position of the maxima in the pattern and plot intensity versus position.

Learning Outcomes

You will be able to:

- Observe the interference pattern formed by monochromatic light passing through a double slit.

- Observe the interference pattern formed by monochromatic light passing through a single slit.

- Analyze both interference patterns to find the minimum and maximum intensities.

- Compare the measurements of both experiments to make some assumptions about light.

Hypothesis

How will the diffraction pattern from a double slit compare to the diffraction pattern from a single slit?

Background

In 1801, Thomas Young obtained evidence of the wave nature of light when he passed light through two closely spaced slits. If light consists of tiny particles (or "corpuscles", as described by Isaac Newton), we might expect to see two bright lines on a screen placed behind the slits. Young observed a series of bright lines and he was able to explain this result as a wave interference phenomenon: the waves leaving the two small slits spread out from the edges of the slits.

In general, the distance between slits is very small compared to the distance from the slits to the screen where the diffraction pattern is observed. The rays from the edges of the slits are essentially parallel.

For two slits, there should be several bright points (or "maxima") of constructive interference on either side of a line that is perpendicular to the point directly between the two slits.

The interference pattern for a single slit is similar to the pattern created by a double slit, but the central maxima is measurably brighter than the maxima on either side. Compared to the double-slit pattern, most of the light intensity is in the central maxima and very little is in the rest of the pattern. The smaller the width of the slit, the more intense the central diffraction maximum.

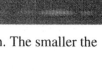

Equation 28.1 relates wavelength (lambda), the number of maxima (m), the slit width (d), and the separation of the maxima in the bright fringes of a diffraction pattern (theta).

$$\sin\theta = \frac{m\lambda}{d}$$

Equation 28.1

For more information see Cutnell & Johnson, Physics, 6th ed., Volume Two, Chapter 27, Section 27.5.

Materials

Equipment Needed	Qty	Equipment Needed	Qty
PASPORT Light Sensor (PS-2106)	1	Basic Optics System (OS-8515)	1
PASPORT Rotary Motion Sensor (PS-2120)	1	Aperture Bracket (OS-8534)	1
USB Link (PS-2100)	2	Linear Translator (OS-8535)	1
Slit Accessories (OS-8523)	1	Diode Laser (OS-8525)	1

Setup

Computer Setup

1. Plug the *USB Links* into the computer's USB ports.

2. Plug the *PASPORT Sensors* into the USB Links. This will automatically launch the PASPORTAL window.

3. Choose the DataStudio configuration file entitled **28 Diffraction of Light CF.ds** and proceed with the following instructions.

SAFETY REMINDER	THINK SAFETY
• Follow the directions for using the equipment.	ACT SAFELY BE SAFE!
• Do not look directly into the laser. It can cause permanent damage to your eyes.	

Equipment Setup

1. Mount the Diode Laser on one end of the Optics Bench. Connect the power supply to the laser.

2. Place the MULTIPLE SLIT SET into the Slit Accessory holder. Mount the Slit Accessory holder in front of the Diode Laser on the bench.

3. Put the rack from the Linear Translator through the slot in the side of the Rotary Motion Sensor. Put the rack clamp onto the rack and tighten its thumbscrew.

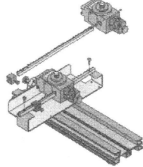

4. Place the rack with the sensor onto the Linear Translator so the back end of the sensor rests on the upright edge of the base of the Linear Translator. Use the thumbscrews to attach the rack to the translator.

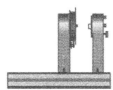

5. Remove the 'O' ring and thumbscrew from the Rotary Motion Sensor pulley so they will not interfere with the Aperture Bracket.

6. Mount the Light Sensor onto the Aperture Bracket by screwing the Aperture Bracket post into the threaded hole on the bottom of the Light Sensor.

7. Put the post into the rod clamp on the end of the Rotary Motion Sensor. Tighten the rod clamp thumbscrew to hold the Aperture Bracket and Light Sensor in place.

8. Place the front end of the Linear Translator at 20 cm. Place the Laser Diode at the 100 cm mark on the bench and the double slit accessory at 90 cm mark on the bench as shown below.

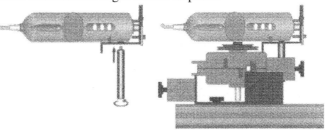

9. Rotate the Aperture Disk on the front of the Aperture Bracket until the number "2" slit is in front of the Light Sensor opening.

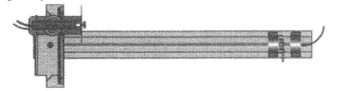

10. Turn on the power switch on the back of the Diode Laser. Adjust the position of the laser and the MULTIPLE SLIT SET on the Slit Accessory so that the laser beam passes through one of the double-slit pairs on the SLIT SET and forms a clear, horizontal diffraction pattern on the white screen of the Aperture Bracket.

11. Record the slit width "a" and slit spacing "d" of the double-slit pattern in meters below.

slit width "a":	
slit spacing "d":	

12.   Move the Rotary Motion Sensor/Light Sensor along the rack until the maximum at one edge of the diffraction pattern is next to the slit in front of the Light Sensor

13.   Place the end stop so it sits up against the Rotary Motion Sensor. (This is very important because you need to start each run from the same position.)

 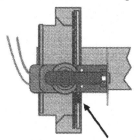 

Record Data

(Hint: Read this all the way through before you begin to take data.)

1.   Click **Start**. Slowly and smoothly, move the Rotary Motion Sensor/Light Sensor so that the white screen on the Aperture Disk moves through the diffraction pattern.

Note: It is very important to turn the pulley on the Rotary Motion Sensor very <u>slowly.</u>

2.   Click **Stop** when you have gone though the entire spectrum.

3.   Replace the double slit accessory with the single slit accessory. Make sure you put it at 90 cm.

4.   Click **Start**. Slowly and smoothly, move the Rotary Motion Sensor/Light Sensor so that the white screen on the Aperture Disk moves through the diffraction pattern.

Note: It is very important to turn the pulley on the Rotary Motion Sensor very slowly.

5.   Click **Stop** when you have gone though the entire spectrum.

Analyze

Data Analysis

• Measure the distance between the central peak and the second maxima on the double-slit graph. Rescale the graph if needed. Use the **Zoom Select** tool to select a region from the central maxima over to the second maxima. **Figure 1**	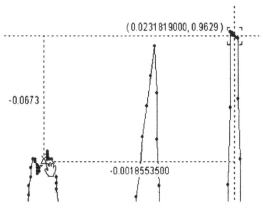
• Click the **Smart Tool** button and move the Smart Tool to the center of the central maxima. Move your cursor to the bottom left corner of the Smart Tool. It will turn into a *delta* symbol. • Drag the delta Smart Tool to the middle of the second maxima. • The delta X is the linear distance between peaks. In the example, delta X is 0.0018553500. **Figure 2** • Record your value below.	(0.0231819000, 0.9629) -0.0673 -0.0018553500

Linear Position, x:

Observations

1. How does the plot of light intensity versus position for the double-slit diffraction pattern compare to the plot of light intensity versus position for the single-slit diffraction pattern?

Calculations

Measure the distance from the double-slit accessory to the aperture bracket and record this as L in meters.

L:

Calculate the angular separation, *theta*, for the second peak using Equation 28.2.

θ:

$$x = L \tan\theta$$

Equation 28.2

Calculate and record your measured wavelength for the laser using Equation 28.1.

m: 2

$$\sin\theta = \frac{m\lambda}{d}$$

Equation 28.1

Remember, d, is the slit spacing value you recorded earlier.

Synthesize

Verification

Compute the percent difference of the wavelengths.   % Difference = $\left|\dfrac{\text{measured - theoretical}}{\text{theoretical}}\right|$ x100%

Insert the theoretical value:

Note: Use the laser's advertised wavelength as the theoretical value.

Percent difference:

Error Analysis

What were the sources of error in this experiment?

Conclusions

How did the diffraction pattern from a double slit compare to the pattern from a single slit?

Do your results support your hypothesis?

Applications

Explain how you would find the specific wavelength of a light source using a diffraction grating?

Extension Problem

The following problem is from Cutnell and Johnson, Physics, 6th ed., Volume Two, Chapter 27, problem 21, page 851.

A single slit has a width of 2.1×10^{-6} m and is used to form a diffraction pattern. Find the angle that locates the second dark fringe when the wavelength of the light is (a) 430 nm and (b) 600 nm?

Activity 29: Spectral Lines
(Light Sensor, Rotary Motion Sensor)

Preface

- *If* you are using the PASCO electronic Workbook specifically designed for this activity, then do the following:
1. Connect the *USB Links* to the computer's USB ports.
2. Connect the *Light Sensor* to one of the USB Links and the *Rotary Motion Sensor* to the other. This will automatically launch the PASPORTAL window.
3. Choose the electronic Workbook entitled: **29 Spectral Lines WB.ds** and follow the directions in the Workbook.

Introduction

Individual atoms emit only specific wavelengths of electromagnetic radiation. The wavelengths, called a spectral pattern, are characteristic of the atom and give clues about its structure. Spectral lines can be viewed and measured with an instrument called the spectrophotometer.

Use a Light Sensor to measure the intensity of the colors in a spectral pattern.

Use a Rotary Motion Sensor to measure the angular position of the colors in spectral pattern.

Calculate the wavelengths of the colors produced by a mercury vapor source based on the measurements of the angular positions of the colors.

Learning Outcomes

You will be able to:

- Determine the wavelengths of the colors in the spectrum of a mercury vapor light.

- Compare a calculated value for the wavelengths with the accepted values.

Hypothesis

How will your measured values for the wavelengths of the colors of light in a spectral pattern compare to the accepted values?

Which color of light in the spectral pattern will have the greatest intensity (brightness) as measured by the light sensor?

Background

An incandescent source such as a hot solid metal filament produces a continuous spectrum of wavelengths. Light produced by an electric discharge in a rarefied gas of a single element contains a limited number of discrete wavelengths - an emission or "bright line" spectrum. The pattern of colors in an emission spectrum is characteristic of the element. The individual colors appear in the shape of "bright lines" because the light that is separated into the spectrum usually passes through a narrow slit illuminated by the light source.

A grating is a piece of transparent material with a large number of equally spaced parallel lines ruled on its surface. The distance between the lines is called the grating line spacing, d.

The parallel lines diffract light that strikes the transparent material. The diffracted light passes through the grating at all angles relative to the original light path. If diffracted rays from adjacent lines on the grating interfere and are in phase, an image of the light source is formed. Rays from adjacent lines will be in phase if the rays differ in path length by an integral number of wavelengths of the light. The first place that an image can be formed is where the path length between two adjacent rays differs by one wavelength, l. However, the difference in path length of adjacent rays also depends on the grating spacing, d, and the angle, theta, at which the rays are diffracted.

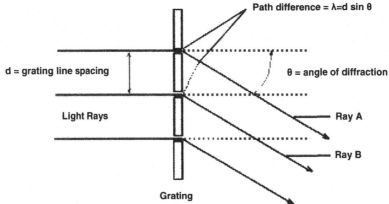

The relationship between the wavelengths of the light, λ, the grating line spacing, d, and the diffraction angle, θ, is as shown in Equation 29.1.

$$\lambda = d \sin \theta \quad \textbf{Equation 29.1}$$

In the diagram above, the path length for Ray A is one wavelength longer than the path length of Ray B.

For more information see Cutnell & Johnson, Physics, 6th ed., Vol. 2, Chapter 27, Section 27.7.

Materials

Equipment Needed	Qty	Equipment Needed	Qty
PASPORT Rotary Motion Sensor (PS-2120)	1	Spectrophotometer Accessory Kit (OS-8537)	1
PASPORT Light Sensor (PS-2115)	1	Mercury Vapor Light Source (OS-9286)	1
USB Link (PS-2100)	2	Basic Optics System (OS-8515)	1
Aperture Bracket (OS-8534)	1	Rod, 45 cm (18") Plated, ½" dia. (ME-8736)	2
Rod Stand Base, Large (ME-8735)	2		

Setup

Computer Setup

1. Plug the *USB Links* into the computer's USB ports.

2. Plug the *PASPORT Sensors* into the USB Links. This will automatically launch the PASPORTAL window.

3. Choose the DataStudio configuration file entitled **29 Spectral Lines CF.ds** and proceed with the following instructions.

Equipment Setup

1. Prepare the Rotary Motion Sensor by removing the thumbscrew, three-step pulley, and rod clamp.

2. Prepare the Spectrophotometer Base by removing the two small thumbscrews and Pinion and by rotating the hinge away from the Base.

3. Attach the Rotary Motion Sensor to the Base hinge with the two small thumbscrews and attach the Pinion to the Rotary Motion Sensor. The screw used to hold the Pinion on the Rotary Motion Sensor should be screwed into the groove of the shaft.

 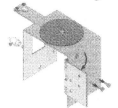 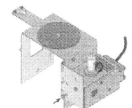

4. Put the Degree Plate/Light Sensor Arm on the Base. Attach the Grating Mount, Light Sensor Mount, and Light Sensor. Position the Focusing Lens.

5. Do not tighten the Grating Mount too tight against the Light Sensor Arm. It should not move as the Light Sensor Arm is rotated.

6. Set the Accessory Bracket disk at the number 1 setting.

7. Put the Spectrophotometer Base onto one end of the Optics Bench. Attach the Rod Stand Mounting Clamps to one side of the Optics Bench.

8. Use support rods and the clamps to raise the bench so that the optical axis of the bench is at the height of the light source.

9. Turn on the light source. Adjust the light source so the beam travels straight down the bench (put a piece of paper behind the grating mount and adjust the light so that the beam is at the center of the mount.)

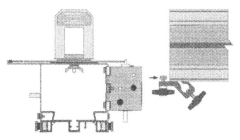

10. Mount the Collimating Slits and Collimating Lens onto the Optics Bench. Adjust the Collimating Slits and Collimating Lens so the image is a constant width along the bench.

11. Mount the grating on the grating holder and adjust the grating vertically.

12. Adjust the focusing lens so the slit image is parallel to the aperture slit.

Note: Adjust the optics bench to align the image vertically if needed.

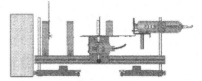

13. If you are using the Mercury Light Source set the Collimating slits to 1.

14. Move the Light Sensor Arm so the Light Sensor is beyond the edge of the first order spectral pattern.

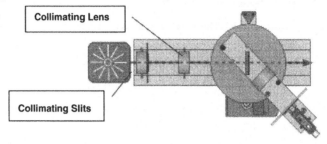

15. You will need to hold the Rotary Motion Sensor against the Degree Base as you slowly rotate the Light Sensor Arm.

16. Record your value for the grating line spacing, d, in meters:

Grating line spacing, d:

Record Data

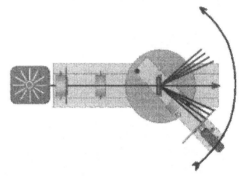

(Hint: Read this all the way through before you begin to take data.)

1. Set the gain on the Light Sensor to maximum (candle).

2. Click **Start**. Scan slowly through the first order spectral lines on both sides.

3. Click **Stop** after you have completed a single run.

Analyze

Data Analysis

You will now find the angle, θ, for each color.

- Click the **Scale to Fit** button. Click the **Smart Tool** button to make the Smart tool appear. Drag the Smart Tool to the top of the first peak at one end of the spectral line pattern. Move the cursor to the bottom corner of the Smart Tool. This will cause a triangle (the "delta" tool) to appear. Drag the "delta" to the peak that corresponds to the same color at the other end of the spectral line pattern.

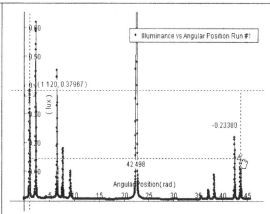

- You will use the change in Angular Position (rad) for your measurement of each color. (In this example, the number for yellow is 42.498.)

- Use the Smart Tool to perform the same analysis for the next pair of peaks in the spectral line pattern and record the values of *theta* below.

- The graph shown will help you identify which lines correspond to which color.

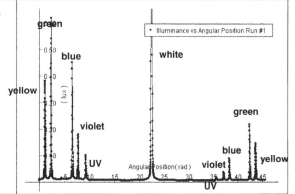

Color	$\Delta\theta(rad)$
Yellow	
Green	
Blue	
Violet	
Ultra-Violet	

Observations

1. What did you observe about the order of the colors in the spectral pattern? Is the order of colors the same as the order of colors in a rainbow, for example?

2. What did you observe about the relative intensity of the colors in the spectral pattern? Were all the colors the same brightness?

Calculations

Divide the angular separation by 120 to find the angle, *theta (θ)*. Use *theta* and the grating line spacing, *d*, to calculate the wavelength, *lambda (λ)*.

$$\lambda = d \sin \theta$$

Color	$\theta = \dfrac{\Delta\theta}{120} (rad)$	λ *(nm)*
Yellow		
Green		
Blue		
Violet		
Ultra-Violet		

Note: You divide the angle by 120 because of the difference between the radius of the degree plate and the radius of the pinion. Be sure your angles are in radians.

Synthesize

Verification

Compute the percent difference for each color by comparing the measured value to the accepted theoretical value.

$$\% \text{ Difference} = \left| \frac{\text{measured - theoretical}}{\text{theorectical}} \right| \times 100\%$$

Color	Yellow	Green	Blue	Violet	Ultra-Violet
Theoretical Wavelength (nm)	578	546.074	435.835	404.656	365.483
Percent difference					

Error Analysis

What were the sources of error in this experiment?

Conclusions

Which color appears to have the highest illuminance (lux)?

Do your results support your hypothesis?

Applications

How could the spectral pattern be used to help identify an unknown light source?

Extension Problem

The following problem is from Cutnell and Johnson, Physics, 6th ed., Volume Two, Chapter 27, problem 39, page 852.

For a wavelength of 420 nm, a diffraction grating produces a bright fringe at an angle of 26°. For an unknown wavelength, the same grating produces a bright fringe at an angle of 41°. In both cases the bright fringes are of the same order m. What is the unknown wavelength?

Activity 30: Photoelectric Effect – Planck's Constant
(Voltage-Current Sensor)

Preface

> • *If* you are using the PASCO electronic Workbook specifically designed for this activity, then do the following:
> 1. Connect the *USB Link* to the computer's USB port.
> 2. Connect the *Voltage-Current Sensor* to the USB Link. This will automatically launch the PASPORTAL window.
> 3. Choose the electronic Workbook entitled: **30 Photoelectric Effect WB.ds** and follow the directions in the Workbook.

Introduction

The purpose of this exploration is to measure the voltage produced when different colors of light fall on a photoelectric tube and to find out the relationship between the frequency of light and the voltage it produces.

Use the Voltage-Current Sensor to measure the voltage produced by the photoelectric effect.

Record and display the voltage produced by each color of light you measure. Use a plot of voltage versus frequency to find the relationship between the voltage and the frequency.

Learning Outcomes

You will be able to:

- Use the photoelectric effect apparatus and the Voltage-Current Sensor to measure the voltages produced by several discrete colors of light.

- Plot a graph of voltage versus frequency for the colors of light that you measure.

- Use the voltage and frequency data to find Planck's constant and compare it to the accepted value.

Hypothesis

What is the relationship between the frequency of light and the energy of light?

Will the graph of frequency versus voltage be linear?

Background

Experimental evidence that light consists of photons comes from a phenomenon called the photoelectric effect, in which electrons are emitted from a metal surface when light shines on it. The electrons are emitted if the light being used has a sufficiently high frequency.

Each color of light in the rainbow has a different amount of energy. Each color also has a certain wavelength. Blue light has a shorter wavelength than red light. If a color of light that has the right amount of energy strikes a metal surface, the light can knock electrons out of the metal.

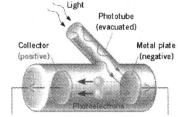

When electrons get knocked out of the metal by photons of light, the metal can produce a voltage. The amount of voltage depends on the number of electrons that get knocked out of the metal. The number of electrons depends on the energy of the light. The energy of the light depends on the wavelength. The color of the light tells you whether the wavelength is long or short.

Albert Einstein knew the relationship between the color of light (or its wavelength) and the amount of energy in the light. He used the relationship between color and energy to explain the photoelectric effect (for which he won the Nobel Prize).

For more information see Cutnell & Johnson, Physics, 6th ed., Vol. 2, Chapter 29, Section 29.3.

Einstein based his explanation on a theory about radiation published in 1901 by Max Planck. In his theory, Planck stated that the emission and absorption of radiation such as light is associated with transitions or jumps between two energy levels. The energy lost or gained is expressed by Equation 30.1 where E equals the energy, "ν" is the frequency of the radiation, and h is a fundamental constant of nature, Planck's constant.

Equation 30.1 $$E = h\nu$$

The kinetic energy of the electron knocked out by the photon depends on the wavelength, or frequency, of the light. Einstein explained the photoelectric effect as shown in Equation 30.2 where KEmax is the maximum kinetic energy of the electrons, and W_o is the energy needed to remove them from the metal (the work function). E is the energy supplied by the photon. When the photon knocks the electron out of the metal, the electron uses a minimum W_o of its energy to escape the cathode, leaving it with a maximum kinetic energy of KE_{max}.

Equation 30.2 $$E = h\nu = KE_{max} + W_o$$

The KE_{max} for an electron is the voltage produced by the metal multiplied by the charge on the electron as shown in Equation 30.3 where V is the voltage and e is the electron's charge.

Equation 30.3 $$KE_{max} = Ve$$

Substituting in Einstein's equation gives Equation 30.4.

Equation 30.4 $$h\nu = Ve + W_o$$

Solving for V (voltage) gives Equation 30.5.

Equation 30.5 $$V = \left(\frac{h}{e}\right)\nu - \frac{W_o}{e}$$

The slope of a graph of voltage (V) and frequency (ν) is the ratio, *h/e*. Multiplying the slope by the charge on an electron (e=1.602 x 10^{-19}) gives Planck's constant, *h*.

Materials

Equipment Needed	Qty	Equipment Needed	Qty
PASPORTVoltage-Current Sensor (PS-2115)	1	h/e Optics and Alignment Kit (AP-9369)	1
USB Link (PS-2100)	1	Mercury Vapor Light Source (OS-9286)	1
h/e Head (Photoelectric Effect) (AP-9368)	1		

Setup

Computer Setup

1. Plug the *USB Link* into the computer's USB port.

2. Plug the *PASPORT Voltage-Current Sensor* into the USB Link. This will automatically launch the PASPORTAL window.

3. Choose the DataStudio configuration file entitled **30 Photoelectric Effect CF.ds** and proceed with the following instructions.

SAFETY REMINDER	THINK SAFETY
• Follow the directions for using the equipment.	**ACT SAFELY**
• Do not touch the mercury vapor light source after it gets hot.	**BE SAFE!**

Equipment Setup

1. Install the two 9 volt batteries that come with the Photoelectric Apparatus into the battery compartment on the apparatus.

2. Put the Photoelectric Apparatus onto the support base assembly.

3. Put the light block into the slot on the back of the mercury vapor light source.

4. Put the coupling bar assembly onto the bottom of the slot on the front of the mercury vapor light source.

5. Put the light aperture assembly in the middle of the slot on the front of the mercury vapor light source.

6. Be sure that the light aperture assembly is lined up with the opening on the front of the light source.

7. Put the lens/grating assembly onto the ends of the horizontal rods on the light aperture assembly.

8. Put the hole at the end of the support base assembly over the pin on the end of the coupling bar assembly.

9. Put the Voltage-Current Sensor banana plugs into the output jacks on the side of the Photoelectric Apparatus.

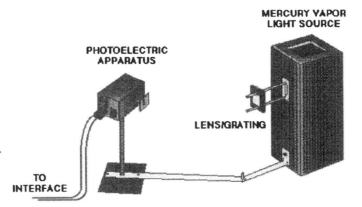

10. Turn on the light source so it has time to warm up before you make measurements.

11. Set the h/e Apparatus directly in front of the Mercury Vapor Light Source. The diffraction grating in the Lens/Grating assembly produces a spectrum pattern to the right and to the left. One will appear brighter than the other due to a feature of the diffraction grating. Move the h/e Apparatus so the brighter spectrum pattern appears on the reflective surface.

12. By sliding the Lens/Grating assembly back and forth on its support rods, focus the light onto the white reflective mask of the h/e Apparatus.

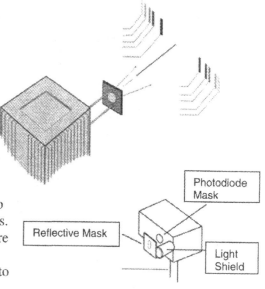

13. Roll the light shield of the Apparatus out of the way to reveal the white photodiode mask inside the Apparatus. Rotate the h/e Apparatus until the image of the aperture is centered on the windows in the photodiode mask. Then tighten the thumbscrew on the base support rod to hold the Apparatus in place.

Record Data

1. Adjust the position of the Photoelectric Apparatus so that the ultraviolet color light from the mercury light source goes through the slot on the front of the apparatus and shines on the photodiode tube inside.

• Note: You will see two violet lines. The line closest to the blue line is the line you want to use for your violet measurement.

2. Turn on the Photoelectric Effect Apparatus and press the PUSH TO ZERO button.

3. Click **Start**. The Start button changes to **Keep**.

4. When the voltage stops going up, click **Keep** to record the voltage in the Table next to the wavelength of violet light.

• The Table below shows the value of voltage for ultraviolet light in the first row.

◇ Data ▾	✕	
☀ Frequency Run #1	✕ Voltage Run #1
Freq (THz)	(V)
820.264	2.009

5. Repeat the process for each of the colors. Note: Use the green and yellow filters for the green and yellow colors of light.

• Note: Open the light shield and check the alignment for each new color. Remember to use the green and yellow filters for the green and yellow colors of light.

• The next section describes how to analyze your graph of voltage and frequency data.

Analyze

Data Analysis

1. Click the **Scale to Fit** button.

2. Click the **Fit** button. Choose **Linear Fit** from the menu.

3. Record the values for Slope and Y-Intercept.

Slope (h/e):

Y-Intercept:

Observations

1. Based on your graph, is the relationship between voltage and frequency direct, inverse, inverse square, or something else?

2. Which color produces the highest voltage?

3. Which color produces the lowest voltage?

Calculations

1. Divide your measured slope by one trillion (to convert from terahertz to hertz) and record the new value.

Slope:

New value:

2. The slope is the ratio of Planck's constant divided by the charge of an electron. Multiply your new value by the charge of an electron. The result is your calculation of Planck's Constant. Record the result.

Charge of an electron: 1.602×10^{-19} C

Planck's Constant (calculated):

3. Calculate and record a percent difference between the measured and theoretical values for Planck's Constant.

$$\% \text{ difference} = \left| \frac{\text{measured-theoretical}}{\text{theoretical}} \right| \times 100$$

Accepted value: 6.626×10^{-34} J s

Percent difference:

Synthesize

Error Analysis

What were the sources of error in this experiment?

Conclusions

Using the equation and your graph, how much work was needed to eject the electron from the metal? (On your graph, the Y-intercept is the work function divided by the charge on the electron.)

Do your results support your hypothesis?

Applications

What products use the photoelectric effect to function?

Extension Problem

In Cutnell & Johnson, Physics, 6th ed., Chapter 29, problem 5, page 900.

Ultraviolet light with a frequency of 3.00×10^{15} Hz strikes a metal surface and ejects electrons that have a maximum kinetic energy of 6.1 eV. What is the work function (in eV) of the metal?

NOTES

NOTES

NOTES

NOTES

NOTES

NOTES

NOTES

NOTES

NOTES

NOTES

NOTES